Advancements in Electric and Hybrid Electric Vehicle Technology

SP-1023

GLOBAL MOBILITY *DATABASE*

All SAE papers, standards, and selected books are abstracted and indexed in the Global Mobility Database.

Published by:
Society of Automotive Engineers, Inc.
400 Commonwealth Drive
Warrendale, PA 15096-0001
USA
Phone: (412) 776-4841
Fax: (412) 776-5760
February 1994

TL
220
.A33
1994

Permission to photocopy for internal or personal use, or the internal or personal use of specific clients, is granted by SAE for libraries and other users registered with the Copyright Clearance Center (CCC), provided that the base fee of $5.00 per article is paid directly to CCC, 222 Rosewood Drive, Danvers, MA 01923. Special requests should be addressed to the SAE Publications Group. 1-56091-475-0/94$5.00.

No part of this publication may be reproduced in any form, in an electronic retrieval system or otherwise, without the prior written permission of the publisher.

ISBN 1-56091-475-0
SAE/SP-94/1023
Library of Congress Catalog Card Number: 93-87534
Copyright 1994 Society of Automotive Engineers, Inc.

Positions and opinions advanced in this paper are those of the author(s) and not necessarily those of SAE. The author is solely responsible for the content of the paper. A process is available by which discussions will be printed with the paper if it is published in SAE Transactions. For permission to publish this paper in full or in part, contact the SAE Publications Group.

Persons wishing to submit papers to be considered for presentation or publication through SAE should send the manuscript or a 300 word abstract of a proposed manuscript to: Secretary, Engineering Meetings Board, SAE.

Printed in USA

PREFACE

Although the major manufacturers have initiated the effort required to take electric vehicle technology from the laboratory through the required development steps to provide a product for the market place to meet the California regulations requiring introduction of "zero emission vehicles" by 1998 (electric vehicles are the prime choice at this time), it is still too early for them to share their efforts publicly. Thus, for 1994 the work offered for publication, SAE special publication <u>Advancements in Electric and Hybrid Electric Vehicle Technology</u> (SP-1023), ranges from information that is immediately of interest to manufacturers "Influence of Battery Characteristics on Traction Drive Performance" and "Thermal Comfort of Electric Vehicles" to longer term interests such as "Chassis Design for a Small Electric City Car" and well into the future with "PEM Fuel Cell Characteristics for EV Application." Also included is some information developed by utilities based on some practical experiences in charging EVs, "Poor Quality Problems at Electric Vehicle's Charging Station" and "Low Fluency Magnetic Field Generated at Electric Vehicle's Charging Station."

We hope that this year's papers will trigger your imagination and provide the foundation for innovative developments that will help the electric vehicle play an important role in our transportation system.

Bradford Bates
Ford Motor Co.

Chairman, Electric Vehicle Committee
Session Organizer

TABLE OF CONTENTS

940293 **Influence of Battery Characteristics on Traction Drive Performance** ...1
    Udo Winter
      Siemens Automotive
    Jurgen Brandes
      Siemens Drive and Standard Products

940294 **Chassis Design for a Small Electric City Car** ...5
    T. G. Chondros, S. M. Michalitsis, and S. D. Panteliou
      University of Patras
    A. D. Dimarogonas
      Washington Univ.

940295 **Thermal Comfort of Electric Vehicles** ...13
    John L. Dauvergne
      Valeo

940296 **Proton Exchange Membrane Fuel Cell Characterization for Electric Vehicle Applications** ...19
    D. H. Swan, B. E. Dickinson, and M. P. Arikara
      University of California, Davis

940297 **Power Quality Problems at Electric Vehicle's Charging Station** ...31
    George G. Karady, Shahin H. Berisha, and Tracy Blake
      Arizona State Univ.
    Ray Hobbs
      Arizona Public Service Co.

940298 **Low Frequency Magnetic Field Generated at Electric Vehicle's Charging Station** ...39
    George G. Karady, Shahin H. Berisha, and Mukund Muralidhar
      Arizona State Univ.
    Ray S. Hobbs
      Arizona Public Service Co.

940336 **Specific Analysis on Electric Vehicle Performance Characteristics with the Aid of Optimization Techniques** ...47
    P. Frantzeskakis, T. Krepec, and S. Sankar
      Concordia Univ.

940337 **The Development and Performance of the AMPhibian Hybrid Electric Vehicle** ...57
    Gregory W. Davis, Gary L. Hodges, and Frank C. Madeka
      United States Naval Academy

940338 **The Selection of Lead-Acid Batteries for Use in Hybrid Electric Vehicles** ...63
    Frank C. Madeka, Gregory W. Davis, and Gary L. Hodges
      United States Naval Academy

940339 **Development of the University of Alberta Entry in the 1993 HEV Challenge** ...69
    M. D. Checkel, V. E. Duckworth, C. B. Collie, and K. M. W. Workun
      University of Alberta

940340   **Hybrid Electric Vehicle Development at the University of California, Davis: The Design of Ground FX** 83
      Rebecca Riley, Mark Duvall, Robert Cobene II, Gregory Eng, Keith Kruetzfeldt, and Andrew A. Frank, Advisor
        University of California, Davis

940510   **Development of a 24 kW Gas Turbine-Driven Generator Set for Hybrid Vehicles** 99
      Robin Mackay
        NoMac Energy Systems, Inc.

940556   **The NGV Challenge - Student Participation, Faculty Involvement, and Costs** 107
      Richard M. Lueptow
        Northwestern Univ.

940557   **Analysis of Data from Electric and Hybrid Electric Vehicle Student Competitions** 113
      Keith B. Wipke
        National Renewable Energy Lab.
      Nicole Hill and Robert P. Larsen
        Argonne National Lab.

940293

Influence of Battery Characteristics on Traction Drive Performance

Udo Winter
Siemens Automotive

Jurgen Brandes
Siemens Drive and Standard Products

ABSTRACT

The development of zero emission vehicles requires new concepts of the electrical drive train. On one hand the vehicle performance for a Californian electric vehicle must give high acceleration values together with enough vehicle range. On the other hand the expected battery performance will be an extreme limitation of capability. In this paper a comparison of an induction motor drive system is given for different working conditions and batteries such as Sodium Sulphur, Sodium Nickel Chloride and Nickel Cadmium.

The results of simulations and practical experience show that exact vehicle predictions can be done if the battery effects such as internal losses and dynamic behaviour are taken into account.

BATTERY CHARACTERISTICS

For analysis of vehicle performance with different batteries a simplified battery model is used. The list below shows the no - load characteristics of single cells:

Pb/PbO2	(lead acid):	2 V
NiCd	(Nicad battery):	1,22 V
NaS	(Sodium Sulphur):	2 V
NaNiCl	(Sodium Nickel Chloride):	2,58 V

To get acceptable energy storage , cells are connected in series and/or in parallel. Higher voltages with Nicad batteries require more serial connections than with NaS or NaNiCl (high-temperature-batteries) and are more difficult for lifetime because of unequal distribution of charge in serial connection. If the cells are connected in parallel equalizing currents have to be controlled.

In serial connections the batteries can be described in the following way:

$$U_{Batt} = U_O - R_i \cdot I_{Batt} \quad (1)$$

The battery voltage drops nearly linearly with current. Typical values are:

	U_O	R_i	
Sodium sulphur	336 V	480 mΩ	120 Ah
Ni-Cd	227 V	210 mΩ	40 Ah
Lead acid	324 V	200 mΩ	50 Ah
NaNiCl	286,4 V	370 - 1300 mΩ	90 Ah

The inner resistance limits the max. power to be taken off out of the battery and is important for the efficiency of the battery.

MAXIMUM POWER OF THE BATTERY

The maximum power of the battery is given with the inner resistance of the battery as follows (see Fig. 1).

$$U_{Batt} \cdot I_{Batt} = \frac{U_O}{2} \cdot \frac{U_O}{2R_i} = \frac{U_O^2}{4R_i} \quad (2)$$

This value is only a theoretical value because working at this point will lead to thermal damage of the batteries. The lifetime of the battery will be reduced drastically.

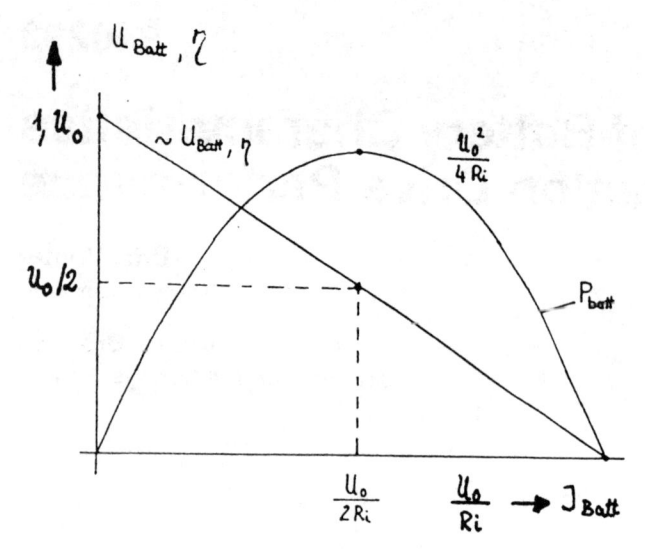

Fig. 1: Simplified (linearized battery model)
From thermal and lifetime considerations the working point for traction drive systems normally is between $2/3\, U_0$ and U_0.

EFFICIENCY OF BATTERY

The efficiency of a battery is given by the ratio of output power to ideal power of battery.

$$\eta_{Batt} = \frac{U_{Batt} \cdot I_{Batt}}{U_0 \cdot I_{Batt}} = \frac{U_{Batt}}{U_0} \quad (3)$$

The efficiency of the battery is increased by using as low power as possible. Therefore the request of a powerful vehicle is automatically reducing the range of this vehicle by higher consumption and worse battery efficiency. For EV-traction drives the battery current is limited by electronic control to get a compromise between power capability and lifetime.

SIMULATION OF TRACTION DRIVE

To work with a battery the traction drive performance should maximize operation with constant power over a wide speed range. The ideal torque, power /.speed curve shows Fig. 2:

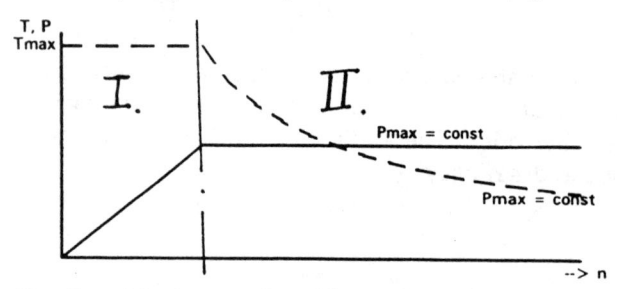

Fig. 2: Ideal curve for drive train performance

INDUCTION MOTOR DRIVE SYSTEM PERFORMANCE

In order to approach the ideal electric drive performance characteristic induction motors are driven with special controls in EV-applications (Fig. 3). In region I (Fig. 2) the drive train is operated with magnetic fields between 0 and a maximum field for control of the torque. The inverter current limit produces the limitation for torque.

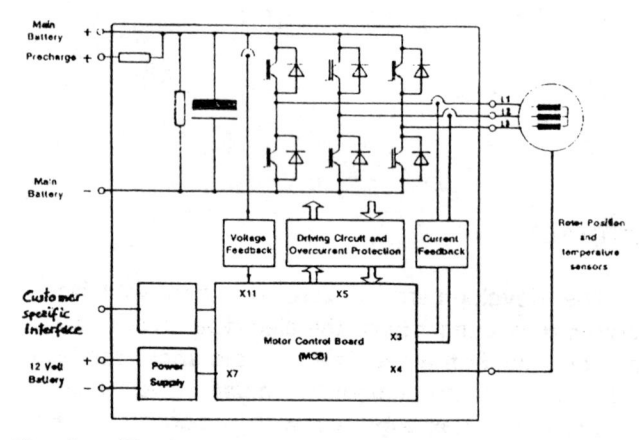

Fig. 3: Typical arrangement of IGBT inverter driven induction motor

Fig. 4 shows the natural performance of an induction motor drive system with constant voltage without a battery current limiting function.

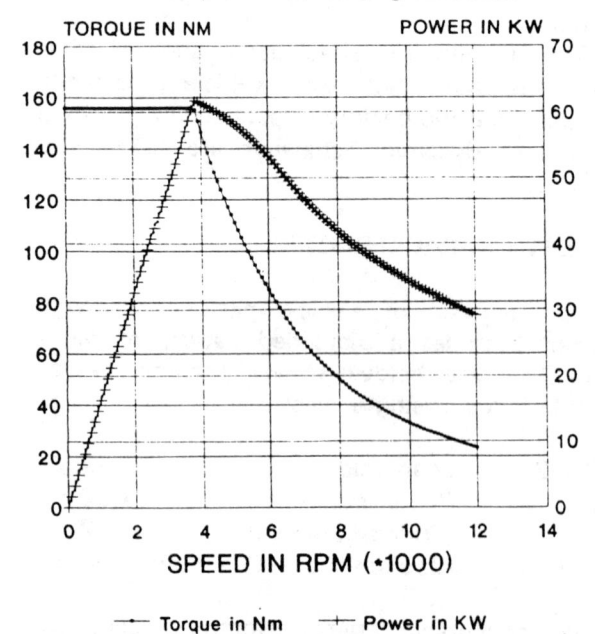

SIEMENS MOTOR 1PV5105-4WS15

Fig. 4: Induction motor drive system at 250 V DC with 400 A max. IGBT inverter (simulation).

In region I the max. torque is given by limited stator (inverter) current I_1 in the following characteristic:

$$T_{max} = 3p \cdot L_h \, Re \, (j \cdot I_\mu^* \cdot I_1) \quad (4)$$

with: $I_\mu = \dfrac{\dfrac{R_2'}{s} + j \cdot 2\pi f_1 L_{\sigma 2}'}{\dfrac{R_2'}{s} + j \cdot 2\pi f_1 (L_{\sigma 2}' + L_h)} \cdot I_1$

- s = slip
- R_2' = rotor resistance referred to stator
- $L_{\sigma 2}'$ = rotor leakage inductance referred to stator
- L_h = main inductance
- I_μ = magnetizing current

In region II the power is not a constant curve because of the limited battery voltage. At constant voltage and high frequency the induction motor cannot produce more torque than given by the stability limit (main wave model for main frequency).

$$T_{max} \approx \dfrac{3p}{2L_k} \cdot \left(\dfrac{(U_{Batt} - 6V)}{2\pi f_1}\right)^2 \quad (5)$$

with:
- f_1: stator frequency, in Hz
- L_k: leakage inductance
- p: number of pole-pairs
- $6V$: Voltage drop at IGBTs

To obtain a powerful drive train the leakage inductance has to be minimized. On the other hand the leakage inductance is limiting the high frequency motor currents (at inverter switching frequency) in the motor and therefore the associated harmonic losses of the drive train.

These harmonic losses are load independent and so they are the most significant losses under partial load condition. Therefore the allowed max. power of a drive train always is a compromise between range and performance.
In region II the power normally is limited by algorithms to adapt the drive train power to the special battery.

INFLUENCE OF VOLTAGE FOR GIVEN MOTOR DESIGN

The voltage is influencing the performance of the drive train. For a given design of the motor the torque at low speed will not be influenced (region I). At higher speeds the torque and power capabilities are depending on the square of voltage.(see formula (5))

In the region of battery current limitation this effect cannot be seen. It is only outside this region that the natural performance (high speed) is occuring.

In the following the drive trains for different electric vehicles are compared. The voltage of NaS, Lead-acid and NaNiCl are close to each other for operation. With NiCd the higher voltage could not be realized on vehicles. For all electric vehicles with different performances the same drive train is analyzed.

SIMULATION WITH DIFFERENT BATTERIES

A. DRIVE TRAIN WITH LEAD ACID BATTERIES

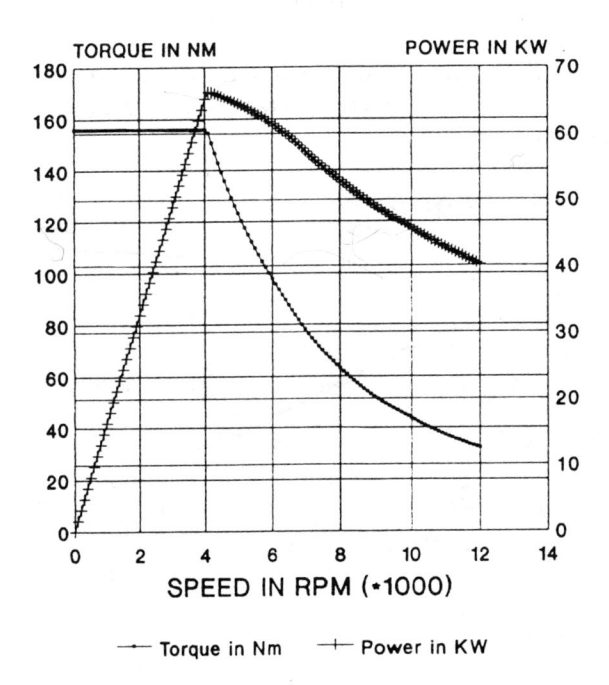

SIEMENS MOTOR 1PV5105-4WS15

Fig. 5: Motor with 324 V batteries and
$U_{Batt} = 324\,V - 0.2\Omega \cdot I_{Batt}$

The power capability with lead acid is very high. As can be seen from the simulation curve, the natural curve of the induction motor is getting stabilized to "constant" power taking inner resistance of the battery into account.

B. DRIVE TRAIN WITH NICAD

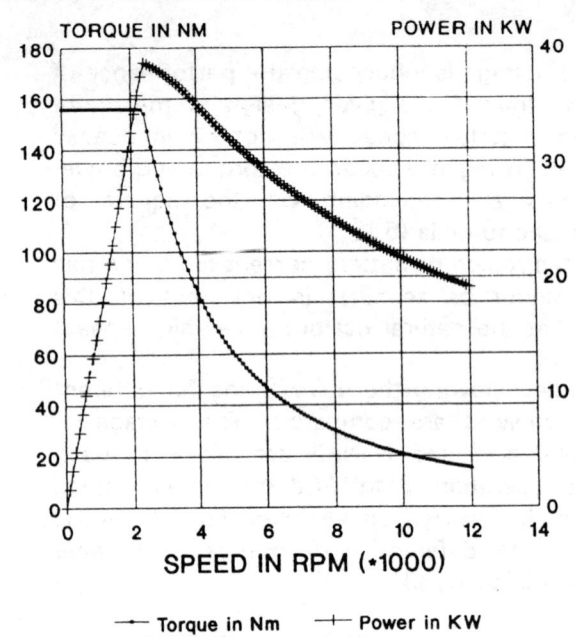

Fig. 6: Motor with NiCad-batteries
$U_{Batt} = 227 V - 0.21\Omega \cdot I_{Batt}$

This drive train was simulated for a small vehicle. With the lower voltage the power capability is reduced.

C. DRIVE TRAIN WITH NAS

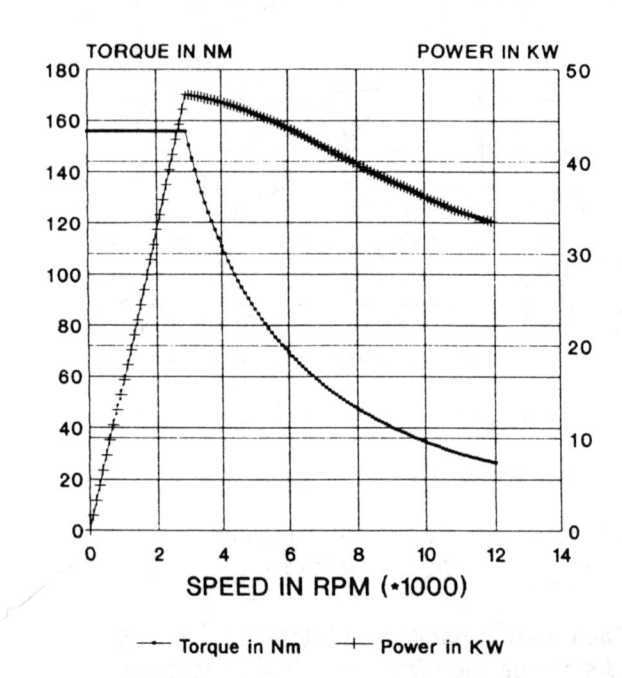

Fig. 7: Drive train with NaS
$U_{Batt} = 336 V - 0.48\Omega \cdot I_{Batt}$

With NaS the vehicle range can be increased. The max. power of about 50 kW leads to a good compromise between power and range within vehicle.

D. DRIVE TRAIN WITH NANICL

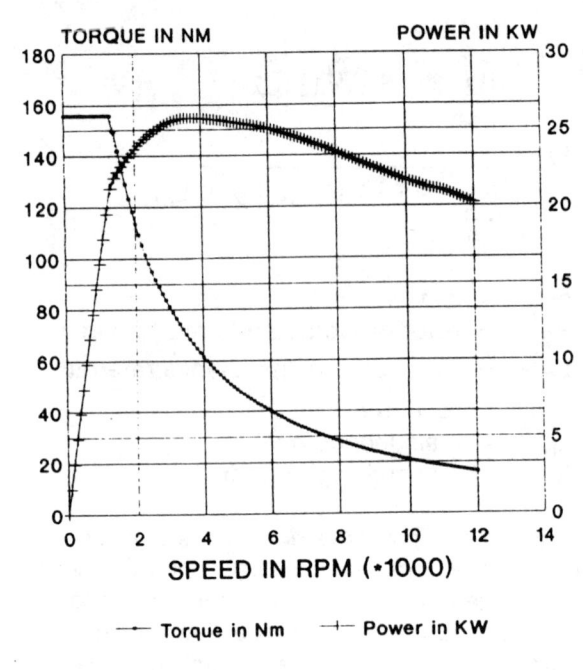

Fig. 8: Drive train with NaNiCl
$U_{Batt} = 286 V - 0.650\Omega \cdot I_{Batt}$
(discharged battery)

The inner resistance of this battery is simulated with 650 mΩ. Under this condition the power is getting nearly constant between 2000 - 12000rpm.

CONCLUSION

Simulation shows the influence on drive train performance of different battery technologies. The inner resistances of NaS and NaNiCl are limiting the possible drive train capabilities for a given battery. These capabilities are theoretically higher for a NiCad and Lead Acid. Considering these conditions the total system efficiency of these lower energy batteries becomes more important. To achieve a good vehicle range with these batteries the power capability of battery has to match the drive train capabillity. Otherwise a good vehicle performance can only be guaranteed for a short range.

940294

Chassis Design for a Small Electric City Car

T. G. Chondros, S. M. Michalitsis, and S. D. Panteliou
University of Patras

A. D. Dimarogonas
Washington Univ.

ABSTRACT

The increasing environmental pollution, the noise emissions, the deteriorating traffic conditions in the big cities, and the decreasing oil deposits have forced the car industries to search for alternative energy sources and new concepts. Increasing R&D efforts in the field of the electric vehicle, which seems to become a substantial part of the fleet of automobiles in the years to come, request for automated and simplified design and simulation techniques. The design of the chassis of E-240, a two-seated electric compact city car, developed at the University of Patras with the use of computer aided software tools, is presented here.

INTRODUCTION

Electric vehicles are once again attracting much attention as an important solution to the problem of improving air quality. Research and development efforts are moving ahead and steady progress is being made [1-4]. Day by day, a lot of progress is being made in the fields of range, acceleration, economy, and also energy and power density and recharging time of the batteries, which are the main drawbacks of the electric vehicles. The process of design and manufacturing, beginning with an idea and ending with a final product, is a closed loop one. The computer-aided analysis capability serves as part of the design process and is also used as a model simulator for the manufactured end product [5,6].

A few years ago, in 1986, an effort began at the Dynamics and Machine Theory Lab of the University of Patras for the design and construction of the E-240, a small electric city car. The intention of this attempt, apart from the main task of the construction and testing of the car, was the development of computer aided design tools. Of course, car manufacturers and designers have devised algorithms, assisted by long experimentation and experience, to support the design of conventional cars.

Such procedures are lengthy and, in general, proprietary. In the case of the design of electric cars, simple and flexible tools have to be introduced since R&D is in quick progress, and numerous researchers and industries are working in the field. The first analysis showed that the ideal vehicle for this purpose should be a small two-seat car with extra space at the back which could be used either as a luggage compartment, or as a power generator space, or even more as an additional seat for children (Fig. 1).

Fig. 1 The artists conception.

Excess weight from the batteries influences the electric vehicle's structure. This extra weight has also an adverse effect on the performance and operating range which can be achieved by the vehicle. There are two obvious choices to increase the performance of an electric vehicle. The first is to install more powerful electric motors, which, however, will not improve the performance due to the demand for heavier batteries. The second option calls for a drastic reduction of the weight of the vehicle. Positioning of the batteries packs above the wheels axles offers serious advantages to the reduction of the frame weight, since these parts of the frame are already strengthened due to the presence of the attachment points for the suspension.

Originally, cars were designed with a relatively heavy chassis frame on which the body rested [7,8]. In modern

vehicles there is no separate frame but it is the body itself that absorbs all forces in spite of lightweight design. In any case, the design and manufacturing process of special vehicles, like small electric cars, has to put some initial criteria for the chassis design. These criteria concerning the chassis design of E-240 were:
1. Low cost equipment and production process facilities.
2. Convenient adaptation to different types of batteries and traction trains.
3. Production of cars with different types of bodies.
4. Simplicity of the assembly and disassembly process.

The above criteria in relation with the need for a flexible manufacturing process, even by small enterprises, led to the selection of an independent frame and a separate body.

DRIVE TRAIN SELECTION

The design of an electric car requires the estimation of its dynamic characteristics in advance. Maximum speed in different road gradients, acceleration, and range have to be determined. The desired driving behaviour of the car predicts the selection of the power-train components. The main components of the power-train are batteries, power generator in the case of a hybrid car, electric motor with controller, and transmission system consisting of gear box, differential and wheels [9,10].

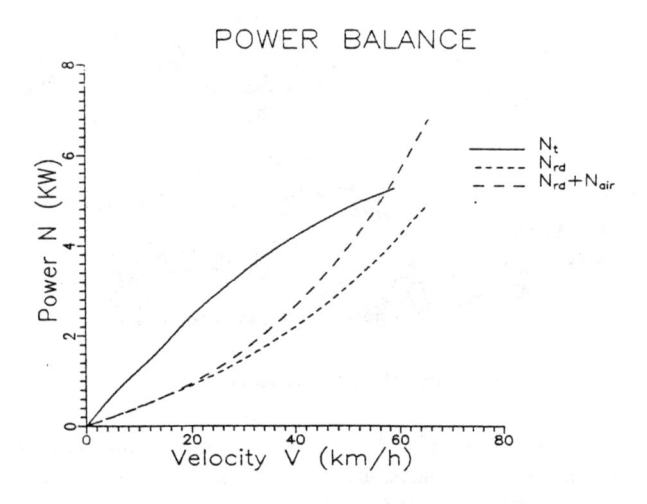

Fig. 2 Power Balance Diagram.

A computer program (DRIVESEL) was developed for the design of the drive-train of the electric car [11]. This software package, simulates the way a designer works to select the drive-line components of an electric or hybrid car and optimizes its function. Additionally, the program simulates the performance of the car with the introduced set of drive-line components, and it is structured in such a way to allow convenient modifications of major vehicle and drive-line parameters. A modular configuration for all drive-line components is used. For gross vehicle weight 900 Kgs, top desired speed 50 Km/h, range between 50 and 100 km and climbing ability 10%, the necessary motor power and torque were calculated. The analysis performed with program DRIVESEL showed that 6 KW power is required to move the car at a speed of 53 Km/h under full load. For 5% road gradient, the top speed of the car under full load is 35 Km/h, while for 10% road gradient, 24 Km/h respectively. Battery capacity of 170 Ah was calculated to achieve the desired performance range. The power balance diagram for uniform car motion in a level road is shown in Fig. 2. N_{rd} and N_{air} is the power necessary to overcome road and air resistance respectively, while N_t is the motor's required power.

Two series - excited 24 V dc SCHABMULLER motors were used, rated at 3 KW each. The two motors are placed transversely under the seats and the power is transmitted to each of the front wheels with the aid of a chain drive running inside the tunnel. This configuration simplifies the drive-line since there is no need for a gear box or differential. Speed reduction ratio is 3.8:1. The motors speed is controlled with the aid of a chopper made by ZAPPI type HFM 400, appropriate for electric cars series DC motors. Voltage rating is 30 - 60 Volts, current maximum 400 A, and frequency of operation 16 KHz. Regenerative braking is activated whenever the acceleration pedal is released or the brake pedal is pressed. In the latter case higher regeneration is achieved. The controller protects the power train by limiting motor current at low speeds, and in the event of excessive motor or controller temperatures.

New generation maintenance-free lead-acid VARTA accumulators have been chosen to power the vehicle. They are high quality traction batteries for load applications. The battery voltage is 48 V with 172 5h/Ah capacity. Lighting is powered by the same batteries through a 48/24 V electronic transformer. These batteries offer 100 Km maximum range to the car.

ERGONOMY

Any modern product that is going to be used by humans must be ergonomically designed. Ergonomy defines functionality and success to the market. Especially, for cars, ergonomy determines not only the functionality but also the living conditions of the driver and the passengers and - most important - safety and driving comfort. Thus, the frame was designed considering comfortable seating space for the passengers and the drive-train components selected during the previous stage. The car was developed from the study of the interior. All interior dimensions were defined with an adjustable front seat in its rearmost normal driving position.

A two dimensional mannequin (Fig. 3) with H-point being positioned at the seating reference point (SgRP) was used [12, 13]. (The H-point is the pivot centre of the torso and thigh on the two-dimensional mannequin). The 95th percentile large male dummy dimensions were used in order to assure comfortable habitability even for tall adults. Design dimensions are shown in Table I. The seat height

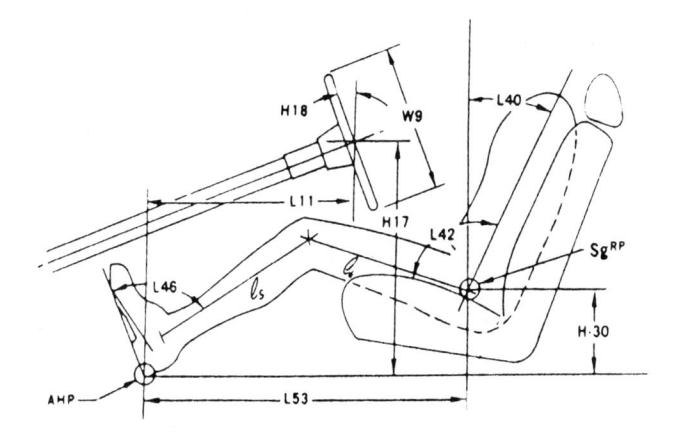

Fig. 3 Vehicle seating configuration.

and also the steering wheel, the pedals and the panels were placed at their normal design position. Steering wheel was positioned with front wheels in straight ahead position. Ergonomic rules were used to form a functional driving position and allow for the design of a comfortable passengers cabin.

TABLE I. Dimensions of E - 240.

Code numbers	Dimensions
L11	360 mm
L40	20° - 25°
L42	100° - 105°
L46	55°
L53	700-850 mm
H17	650 mm
H18	40°
H30	240 mm
W9	360 mm
l_s	444 mm
l_t	456 mm

Actually, the exterior form had to be the outline of the spaces designated for passengers, luggage, mechanical and electrical components. For the final determination of the driver's seat position, the inner boundaries of the floor, the engine compartment, and the roof were drawn. The location of the front wheels was designed so as not to deteriorate the pedals location conditions and the seating comfort of the passengers. The rear wheels were placed at the rear end of the car so that the wheel arches should not decrease the space left for the luggage compartment. Fig. 4 shows the main components layout.

STATIC ANALYSIS

A modern vehicle is, throughout its life, subjected to a variety of random and repetitive input loads and vibrations. These may be applied internally or externally, either way the frame and body must be strong enough to withstand them without serious structural failure [7, 8]. In cars with an independent chassis the frame serves as a carcass to which the motors, the drive-line components, the non suspended mechanical parts, and the body are fastened. The body, the drive-line, the wheels, the suspension and braking system, interact with each other closely. This interaction determines the car's road performance, active safety and driving comfort. For this purpose the frame should exhibit sufficient rigidity and stiffness so that the relative disposition of the installed mechanisms remains unchanged, and the deformations of the body are kept to a minimum under the action of inertia and reaction loads.

The location of the accumulators of the vehicle was a major problem because maximum space had to be left for the mechanical parts, the passengers and the luggage. Batteries location influences seriously the car's road performance. There were two possible solutions for the batteries' storage. The first was to put them under the passengers' cabin and place the motors in the front compartment of the vehicle. With this arrangement the mass should be concentrated to the centre of the vehicle yielding better road performance. This option inevitably affects the height of the car and deteriorates its aesthetics. The second option was to put the batteries in the front and the rear part of the car, and place the motors under the seats as shown in Fig. 4. This arrangement offers better safety to the passengers, since the accumulators are isolated from the cabin, at the extremes of the car. The transverse elements bearing the batteries were designed as independent sub frames also supporting the suspension elements. Thus the batteries' weight, 36% of the total car's laden weight, is not acting on the main frame or the body, providing light construction and good rigidity characteristics.

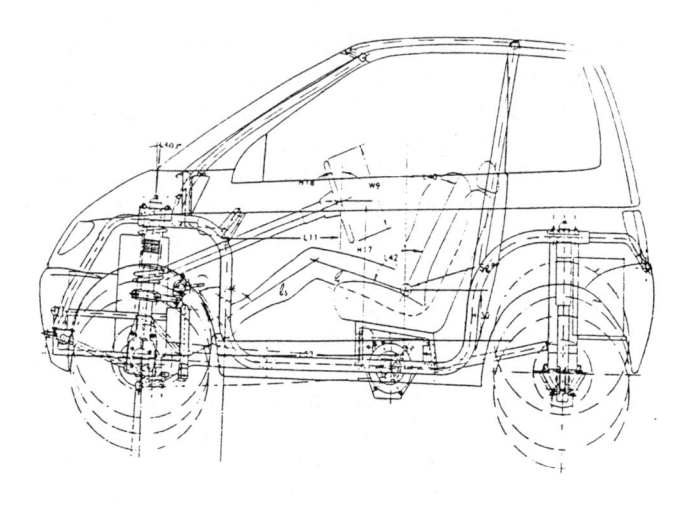

Fig. 4. General arrangement of E - 240.

The main loads acting on the frame are those caused by the body, the accumulators, the electric motors, the driver and passengers and the luggage. These weights are:

1. 280 Kg for the accumulators.
   (164 Kg in the front part of the vehicle and 116 in the back).
2. 50 Kg for the two electric motors.
3. 35 Kg for the motors' controller and the wiring.
4. 150 kg for the passengers
5. 50 Kg for the luggage.
6. 250 Kg for the body uniformly distributed along the frame.

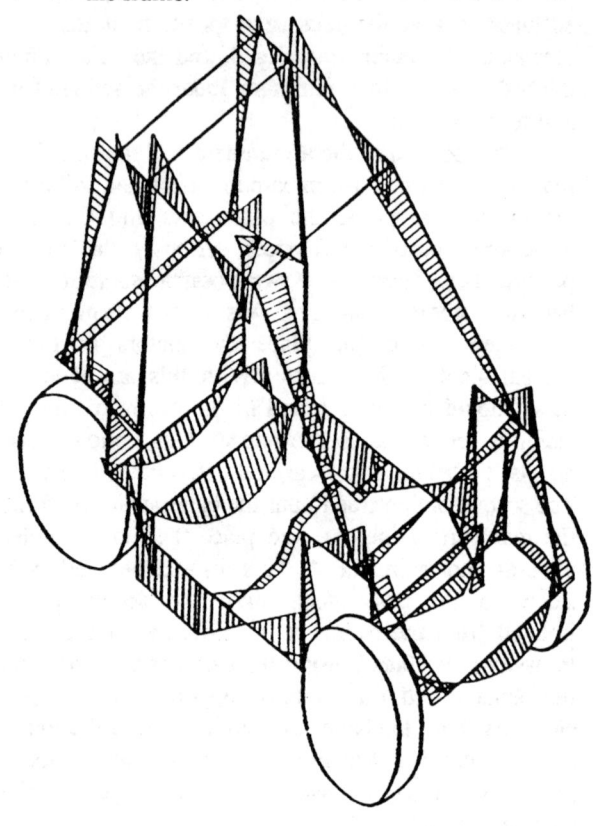

Fig. 5. Loading of the frame members.

All the above loadings, except for the body, were considered concentrated at the points of application. This loading does not include of course, the suspension, tires and brakes' weight, since they are not acting on the frame.

After an initial static analysis, the general requirements for the chassis were determined. The dimensions available during the ergonomy study gave way to the design of the main chassis elements and lay out. Static analysis using a finite elements method was performed for the determination of forces and moments acting on the frame members. Fig. 5 shows loading conditions acting on each member. From this analysis the tubes external diameters were selected accordingly. External diameters for the tubes range from 27 to 42 mm. The frame weight is 40 Kg, while the unladen weight of the car is 730 Kg.

DYNAMIC ANALYSIS

A simplified dynamic analysis followed the static analysis in order to take into consideration the dynamic loading of the frame caused by the vertical motion of the car. This analysis was necessary to assure that the frame would not fail under the action of the dynamic loading which appears when the vehicle is running on a rough road. The same analysis gave also the suspension's springs and dampers constants.

A vehicle motion is characterised by continuous variation of the forces acting due to wheel-road interaction. These variations depend on shape and dimensions of road irregularities and inertial and elastic characteristics of vehicle parts. A reduced vibrating system for determining the variable loads that act on a two-axle vehicle running on a road will be used [14, 15]. Calculations for a two axle automobile are performed independently assuming that the sprung mass is divided in two parts, $m_{s.f}$ for the front and $m_{s.r}$ for the rear suspension (Fig. 6).

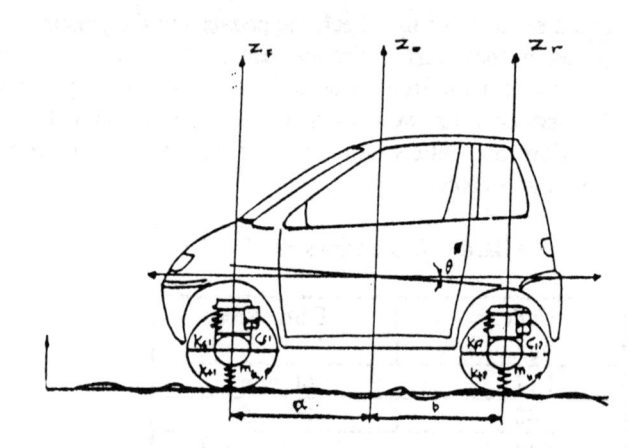

Fig. 6 Car model for dynamic analysis.

Moments equation around the centre of gravity yield:

$$m_{s.f} = m_s [b / (a + b)] \qquad (1)$$

$$m_{s.r} = m_s [a / (a + b)] \qquad (2)$$

where m_s is the total vehicle sprung mass and a, b are the coordinates of the vehicle centre of gravity.

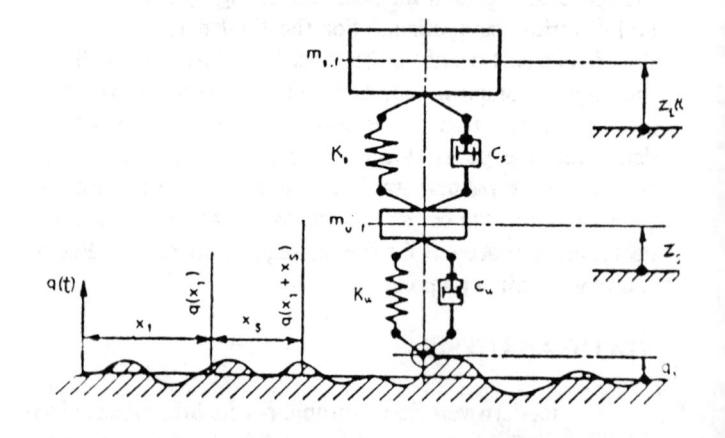

Fig. 7 Two-degrees of freedom system.

Assuming that the wheels do not break away from the road surface, the differential equations of the equivalent vibrating system for the front axle (Fig. 7) can be written as

$$m_{s.f} z_1 + c_{s.f}(z_1 - z_2) + k_{s.f}(z_1 - z_2) = 0 \quad (3)$$

$$m_{u.f} z_2 + c_{u.f} z_1 + c_{s.f}(z_2 - z_1) + k_{s.f}(z_2 - z_1) + k_{u.f} z_2 = f(t) \quad (4)$$

where $m_{s.f}$ is the front axle unsprung mass, $k_{s.f}$, $k_{u.f}$ are the spring constants for the front spring and tyre, and $c_{s.f}$, $c_{u.f}$ are the damping coefficient for the front damper and tyre. A similar system of equations is solved for the rear axle. The function f(t) in Eq. (4) represents the vertical force caused by road disturbance. This force is given by the equation [6];

$$f(t) = k_{u.f} q(t) + c_{u.f} q(t) \quad (5)$$

where q(t) is the variation of the road displacement. Assuming that q(t) has the form:

$$q(t) = q_0 \sin(\omega t)$$

where q_0 is the maximum amplitude and ω is the circular frequency of ground disturbance, the expression for f(t) takes the form;

$$f(t) = k_{u.f} q_0 \sin(\omega t) + c_{u.f} q_0 \omega \cos(\omega t) \quad (6)$$

Equations (3), (4) and (6) represent the unconstrained mathematical model which is known as the "jumping" model [8, 14]. These equations are solved in time domain with the aid of a RUNGE-KUTTA algorithm. Input data shown in Table II include the car and components weight and estimated initial values for front and rear suspension characteristics. Progressive elimination in $k_{s.f}$ and $k_{s.r}$ with a respective increase in $c_{s.f}$ and $c_{s.r}$ values gave the set of the corrected values shown in Table III.

TABLE II Input data for the dynamic analysis

Front Suspension	Rear Suspension
$m_{s\,f}$ = 415 kg	$m_{s\,r}$ = 435 kg
$m_{u\,f}$ = 32 kg	$m_{u\,r}$ = 35 kg
$k_{s\,f}$ = 120000 Nt/m	$k_{s\,r}$ = 150000 Nt/m
$k_{u\,f}$ = 350000 Nt/m	$k_{u\,r}$ = 350000 Nt/m
$c_{s\,f}$ = 500 Nt/m/sec	$c_{s\,r}$ = 500 Nt/m/sec
$c_{u\,f}$ = 4500 Nt/m/sec	$c_{u\,r}$ = 4500 Nt/m/sec

TABLE III Output data from the dynamic analysis.

Front suspension	Rear suspension
$k_{s\,f}$ = 78400 Nt/m	$k_{s\,r}$ = 96200 Nt/m
$k_{u\,f}$ = 350000 Nt/m	$k_{u\,r}$ = 350000 Nt/m
$c_{s\,f}$ = 3500 Nt/m/sec	$c_{s\,r}$ = 3500 Nt/m/sec
$c_{u\,f}$ = 4500 Nt/m/sec	$c_{u\,r}$ = 3500 Nt/m/sec

The "jumping" model for each axle is a two degrees of freedom system (Fig. 7). Thus, it is characterised by two natural frequencies f_{us} and f_s for the unsprung and sprung masses of the vehicle. Assuming the motion of unsprung mass as the disturbance for the sprung mass, the ratio of the corresponding frequencies must satisfy the condition

$$f_{us}/f_s \rangle 1.41 \quad (7)$$

in order to reduce transmissibility to a minimum [14]. Using the selected values for the front and rear suspension characteristics these ratios become

$$f_{us.f}/f_{s.f} = 5.5 \qquad f_{us.r}/f_{s.r} = 3.5$$

which satisfy the aforementioned condition, particularly for the front axle where heavier loading occurs.

THE FRAME

After the completion of the ergonomic study, the selection of the drive-line components, static and dynamic analysis, most of the parameters necessary for the frame design were determined. Details concerning the assembly of various components on the frame were considered too. To start with the frame design some initial requirements and specifications had to be set:
- Sufficient strength, rigidity and stiffness of the frame. This is essential for the safety of the driver and the passengers. The extra concentrated weight of the lead-acid batteries complicates the problem.
- Light construction. Electric vehicles are heavy mostly because of the accumulators, so the unladen weight must be reduced in order to increase the power-weight ratio.
- Low production cost. One of the electric vehicles' disadvantages is their high cost compared to conventional cars. Thus, the cost has to be decreased by taking advantage of a well-designed, lightweight and simple construction.
- Functionality. Due to the small size of the vehicle, many mechanical parts have to be assembled and interact with each other in a specific small volume. The volume handling is very important for the car's road performance, its safety, its driving comfort and the convenient service.
- Easy assembly and provisions considered for the frame to provide space for a small internal combustion engine, for the hybrid version, and also, to bear different types of bodies.
- Development of disassembly techniques that would facilitate recycling of various parts and components.

These parameters led to the design of a truss structure type chassis, and also the use of lightweight mechanical parts. Brakes were calculated using available software tools [5, 16]. The car is equipped with four drum type, dual hydraulic circuit, brakes. The drum diameter is 185 mm. The brake pump is located vertically to save space for the front batteries and progressive braking of the rear wheels is achieved with the aid of a pressure distribution system. The steering is based on the FIAT PANDA system with rack and pinion mechanism. The wheel diameter is 360 mm and 3.25 turns from side to side are required. With this steering mechanism the turning circle diameter is 8720 mm.

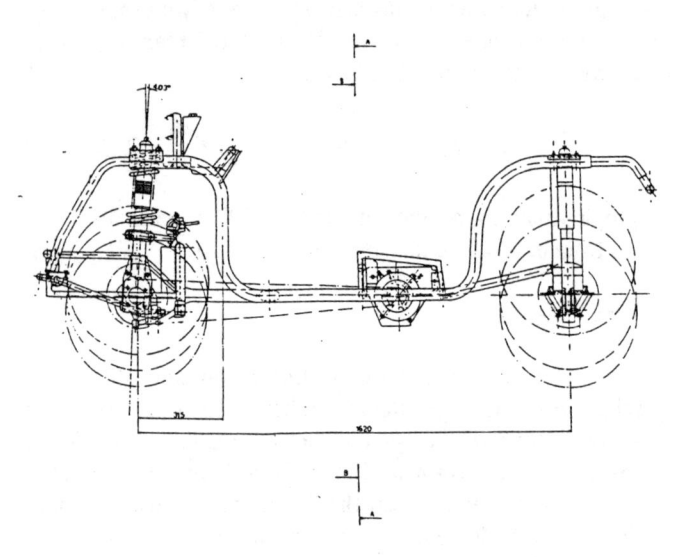

Fig. 8 Side view of the frame.

Figures 8 and 9 show the side and top view of the frame. The designed frame is a combination of a perimeter and a ladder-type frame. There are two side members in the height of the car's floor which rise to pass above the front and rear wheels' arches. The side members are connected together with several cross-members which give to the frame the necessary torsional stiffness. Two of them are placed under the floor so that they can bear the passenger's and the motors' weight.

A cross member comes under the control panel in front of the passengers' cabin and helps to protect them in the case of an accident and also, it supports the console, the steering wheel and the electronic equipment of the vehicle. A transverse element placed between the front "engine" compartment and the passengers' compartment protects the passengers' knees and feet, and bears the upper ends of the McPherson struts. Finally, two more transverse elements "close" the frame in the front and the rear part, constituting the windshield perimeter and a roll bar over the passengers heads completes the frame structure. These members are bolt fitted on the main frame through special attachments. The Ω type rising and lowering of the main side members at the wheels arches region in combination with the rigid transverse members provides good crushing characteristics to the ends of the vehicle.

The frame was constructed in four main stages. In the first stage the individual side and cross tubular members were cut and bent properly. An automatic tube bending machine was used to form the frame parts. All members

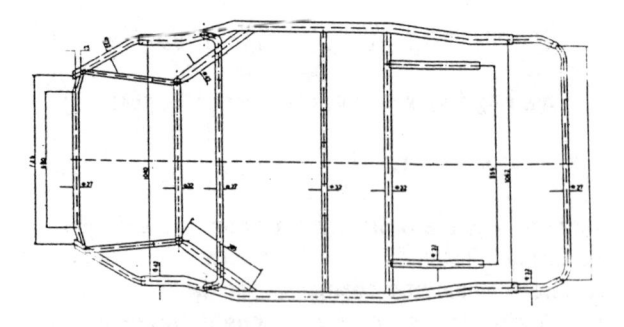

Fig. 9. Top view of the frame.

were placed in a special bed taking the correct position and inclination, and were welded. During the second stage the front and rear suspensions steel sub frames were constructed. These sub frames are also used to bear the batteries weight. An advantage is that they are independent from the main frame and can be easily removed. Also, at this stage the supports of all the mechanical parts (such as the steering system, the gears, the electric motors, the brake pump, etc.) were constructed. At the third stage the individual mechanical parts' supports were placed on the frame and welded. All the weldings were checked and the whole construction was covered with a rust-resisting material and painted. By the end of this stage the frame was ready for the installation of the mechanical parts (Fig. 10).

Fig. 10. The frame of E - 240.

The fourth stage comprises the assembly of the chassis mechanical components. The front and rear suspension assemblies, the brakes, the wheels, the steering system, the electric motors and the transmission units were placed on the special supports. For the testing measurements and the evaluation of the car's electric and mechanical components an open type steel body was constructed and placed on the frame (Fig. 11). Batteries were put on the vehicle after the completion of the body assembly

Fig. 11. The testing prototype.

CONCLUSIONS

Since R&D in electric cars is in a quick progress and changes and modifications in the arrangement and functionality of their mechanical and electrical parts are everyday practice, a simplified tool is necessary for the chassis design. The outlined procedure provides a complete solution to the problem of the design of the chassis for a small electric car. First it provides an initial estimation of the frame elements through a finite element static analysis and, furthermore, through dynamic analysis it evaluates the total applied loads during the car motion, thus facilitating the suspension elements selection. Changes in the design parameters can be easily introduced in the computer programs. The programs can be expanded to perform optimisation techniques, in order to cope with financial criteria that are important in the process of the industrial production of the car.

REFERENCES

[1] Society of Automotive Engineers, Inc., 1992, *Electric and Hybrid Vehicle Technology*, No. SP-915, USA.
[2] Society of Automotive Engineers, Inc., 1991, *Electric Vehicle, Design and Development*, SP - 862.
[3] Maggeto G. 1992, *Advanced Electric Drive Systems for Buses, Vans and Passenger Cars to Reduce Pollution*, Synthesis Report, AVERE.
[4] Seiffert U., P. Walzer, 1990, *Automobile Technology of the Future*, Society of Automotive Engineers, Inc., USA.
[5] Dimarogonas A.D., 1989, *Computer Aided Machine Design*, Prentice Hall, UK ,
[6] Beam W. R., 1990, *Systems Engineering, Architecture and Design*, Mc Graw Hill USA.
[7] Giles, J.G., 1971, *Body Construction and Design*, Iliffe, London.
[8] Lukin P., G. Gasparyants, V. Rodionov, 1989, *Automobile Chassis, Design and Calculations*, MIR Publishers, Moscow.
[9] Wong J. Y., 1978, *Theory of Ground Vehicles*, Wiley Interscience, N.Y.
[10] Gott G. G., 1991, *Changing Gears, The Development of the Automotive Transmission, SAE Inc.*
[11] Chondros T. G., Krouskas S., Dokos D., Dimarogonas A.D., "Computer Aided Selection of Drive-Trains for Electric Cars", 1992, Submitted for publication.
[12] Society of Automotive Engineers, Inc., 1991, *SAE Handbook, Volumes 1-4*, USA.
[13] Society of Automotive Engineers, Inc., 1980, *Devices for Use in Defining and Measuring Vehicle Seating Accommodation*, num. J826, USA.
[14] Dimarogonas A.D., 1976, *Vibration Engineering*, West Publishing Company, St. Paul, USA.
[15] Artamonov M.D., V.A. Ilarionov and M.M. Morin, 1976, *Motor Vehicles, Fundamentals and Design*, MIR Publishers, Moscow.
[16] Dimarogonas A.D., 1992, *MELAB, Computer Programs for Mechanical Engineers*, Prentice Hall, Englewood Cliffs, NJ, USA.

940295

Thermal Comfort of Electric Vehicles

John L. Dauvergne
Valeo

ABSTRACT

In a onventionnal vehicle, it is easy to find thermal energy to be supplied for the thermal comfort in the passenger compartment. That is not the case for the electric vehicles for which this paper proposes technical solutions.

INTRODUCTION

Everybody commonly agrees that electric vehicles will be slower than conventional ones ; they will have also less autonomy. It is less obvious that they will be differentiated by new problems of source of energy for the climatic comfort, i.e for both heating and air-conditioning. These problems will also appear for hybrid vehicles, during phases in "electric mode only".

At first, it is necessary to keep in mind some <u>orders of magnitude</u> about small European or Japanese cars, which have about the same dimensions as future electric or hybrid vehicles designed mainly for urban use. The thermal power as installed in this type of vehicle is dimensioned to provide an acceptable level of comfort with external extreme conditions, i.e :
- in winter :   - 18°C,
- in summer : + 45°C, with a 1000 W/m² solar load,

and this, within "convergence" lead-times as short as possible. Practically, the power of the heater core and of the air-conditioning evaporator are 4 to 6 kW approximately.

Such powers are available from the engine cooling circuit for the heater, and from the compressor for air-conditioning (which absorbs 2 to 3 kW when idling, producing twice more frigorific power at the evaporator).

The conditions of comfort usually obtained are indicated in table I.

Table I - Conditions of comfort

	Heating Winter	Air conditioning Summer
External conditions	- 18°C	+ 45°C 1000 W/m²
Blown air temp.	+ 35°C mini	+ 7°C maxi
Temperature inside vehicle	+ 25°C at feet	21°C at heads
Installed power	4 kW	6 kW

The thermal power is carried by the air stream from the blower. Practically, the amount of air going through the passenger compartment is 40.10^{-3} m³/s but the system is dimensioned for 110.10^{-3} m³/s.

This thermal power is spent (fig. 1) for one half by dissipation through the walls of the passenger compartment and for one half with the air going out of the vehicle.

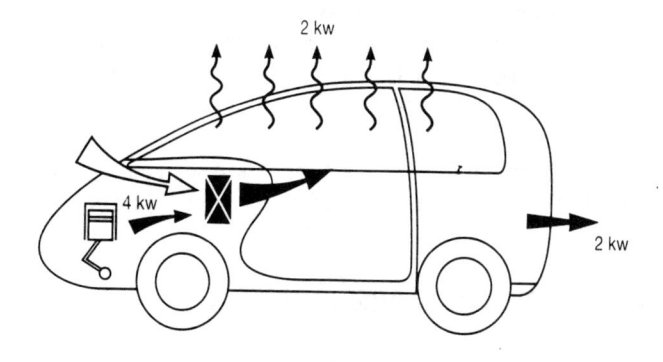

Fig1. Balance of energy for a conventional vehicle

It must be reminded that the heating system is also used for demisting and defrosting of the windshield (regulation FM's 101, ECE 78/316) and of lateral glazing.

SOURCES OF ENERGY : THE CASE OF ELECTRIC VEHICLES

Electric vehicles will not have sufficient energy sources for climatic comfort system :

- the heat rejections of the traction chain exist , however they are dispersed in the vehicle (electric motor, batteries, electronic unit for power control) ; their level is generally low (no more than 2 kW at peaks), but above all it appears randomly (it is variable according to the mission profile, with no heat rejection at stops, and with a maximum during regenerative braking). Nevertheless, they must be used for heating.

- It is not realistic to have the air-conditioning compressor driven by the traction electric motor: it does not run during the traffic stops (precisely when the maximum cooling power is required).
Even for the hybrid vehicle, these conclusions are valid, due to the necessary "electric mode only".
It is clear that it is necessary to develop technical solutions using "on board stored energy":
  - either from the traction batteries
  - or from a specific storage
In both cases, it will be necessary to design the climatic comfort system to correctly use the energy, in order to save autonomy and weight :
  - reduction of thermal losses of the vehicle
  - optimization of the use of energy

REDUCTION OF THERMAL LOSSES

It is necessary to limit the heat transfer through the walls of the passenger compartment :
- Air leakage at door-seals
- Wall insulation (floor, doors, roof) : the use of plastic body panels will contribute to reduction of heat transfer.

Another concern is related to the glazing. The "greenhouse effect" can be minimized with athermic glazing.

OPTIMIZATION OF USE OF ENERGY

Several optimization ways are possible to enable to reduce power consumption during the use of the vehicle.

1. <u>Anticipated operation of heating and air-conditioning</u>

Pre-heating and pre-conditioning, working when the vehicle is "plugged" on the battery charging station, enable to achieve two objectives :

a) - to reach a temperature inside the vehicle at in a level that it will be possible to eliminate the "convergence" period ; when the vehicle is use, it is only necessary to spent the energy corresponding to the heat losses, in order to keep the temperature at an acceptable level. Practically, the equipment will be dimensioned to obtain, after 20 to 30 minutes : 25°C for heating in winter, and 21°C for air-conditioning in summer.
However, the electric power consumption will be limited to 2 kW from the battery charging station.
It could be ideal to "plug" the vehicle each time that it is parked.

b) - to get climatic comfort conditions as soon as the user enters in the vehicle.
That is an advantage of the electric vehicle !

2. <u>Circulation of air inside the passenger compartment</u>

- It is possible to reduce thermal power by avoiding heat rejection with the air going out from the vehicle (fig 2), for instance by recirculation.
The ratio of recirculation can be 100 % in air-conditioning, due to the fact that urban travels are generally short enough.

For heating, it will be limited to 75 % (25 % fresh air) in order to eliminate water vapor from passengers and to avoid mist formation.

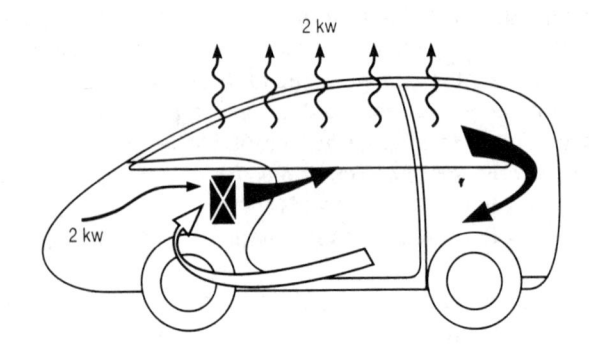

Fig 2. Effect of Recirculation in an electric vehicle

Under these conditions, it is possible to limit energy consumption for climatic comfort only to the heat losses through the body walls.

- The distribution of air must be optimized (fig 3,4). It is recommended to reduce the cold-wall effect by an air layer all along the body walls and glazing (to avoid mist formation as well).

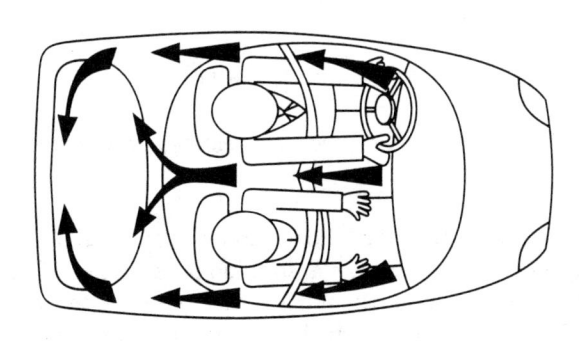

Fig 3. 4 - Optimized distribution of air

3. <u>Auxiliary devices</u> :
- The windshield and the rear glass with integrated electric heating make possible the optimization of electric consumption.
A 400 watt glass-heating with a time is generally sufficient.

- It is also necessary to reduce the power consumption of the blower to 150 Watt (absorbed) for 50 Watt in normal use. The condenser cooling blower will absorb another 150 Watt (it will have to cool also the compressor).

THE TECHNICAL SOLUTIONS

1. Pre-heating burners (using petrol or gas) which are used for direct air or for indirect water heating are very efficient and easy to use with a timer or a remote control. They have the advantage of a very compact energy storage, however they are not totally "zero-emission".

2. Electric solutions supplied by the main traction battery.
   - Electric resistors :

The use of electric resistors from the traction battery create several problems of control of absorbed electric power and of maximum temperature (risk for plastic parts).

The PTC (Positive Temperature Coefficient) resistors enable to solve these two problems (fig.5). Their natural working mode at a constant temperature enables to realize, by adjusting the air flow going through, a self adjustment of the power consumption according to the needs in thermal power.

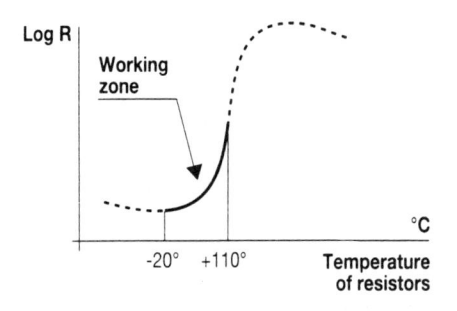

Fig 5. PTC typical curve

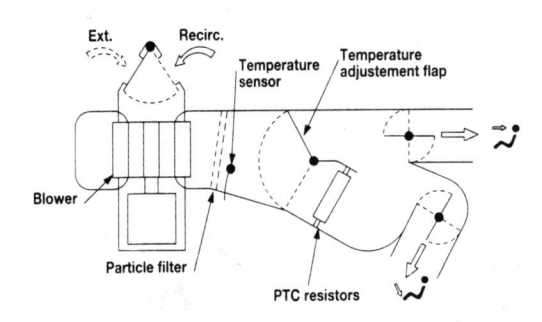

Fig 6. Heater with PTC resistor

Fig 6 shows the principle of a heater using PTC resistors. The temperature adjustment flap is controlled by the user to get the correct amount of air through the resistors.
Two levels of power are generally designed :
    - 2 kW approx. for pre-heating, when the vehicle is "plugged"
    - 1 kW approx. for heating (to keep thermal balance), when it is in use.

The transition temperature is chosen to limit the maximum temperature according to plastic parts near to the resistors.

- Air-conditioning

The installation consists of a classical A/C loop (condenser, evaporator) with an electrically driven compressor ; the electrical motor is supplied from the main traction battery through a power electronic control unit.

It can be either a synchronous motor with a DC/AC converter, or a permanent magnet with a variator.
In both cases, the frigorific power can be adjusted by variation of speed of the electrically driven compressor. Again, two different levels are chosen :
- Pre-cooling with a power of 2 kW (giving 5 kW approx. at evaporator)
- Compensation of heat load when the vehicle is used with 1 kW (2,5 kW approx. at evaporator.
- Reversible air-conditioning for heating (heat pump) (fig. 7).

The operation of the A/C loop as a heat-pump enables to get the best efficiency from the electrical power from the main traction battery : on the condenser (i.e the internal heat exchanger), the thermal power is higher than the one given to the compressor. This kind of system, which is very efficient at medium outside temperatures (more than + 5°C), needs an additional heater (such as PTC resistor) for low temperatures due to the low energy content of outside air and to the risk of icing of the evaporator.

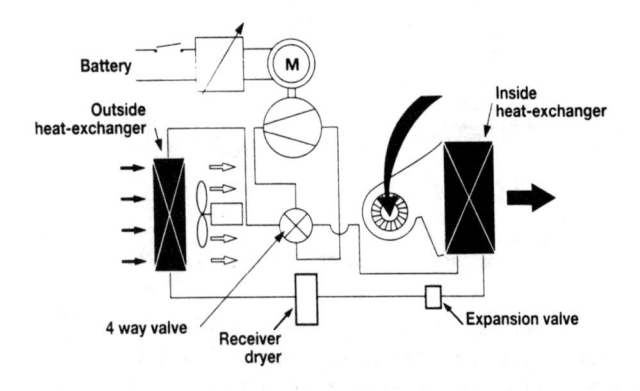

The only devices to be considered are the ones for which the ratio.:

$$\frac{\text{Stored Energy(J)}}{\text{Weight(storage + restitution device(kg)}}$$

is better than the one of the batteries.

Amongst these systems, we find :
- Latent heat storage :
An agent can be melted with storage of thermal energy ; the restitution through a water/glycol circuit can be used for heating.
- Solid sorption systems, which present the interest of a storage of energy for heating, air-conditioning and pre-heating.

The fig 8 shows the principle of the installation : the reactive solid R located in the reactor picks up the saturating vapor V from the liquid of the evaporator, according to the reaction:

$$R + V \longrightarrow (RV) + Q$$

The heat Q is either used for heating, or evacuated outside.

Simultaneously, the consumption of vapor V evaporates the liquid, with absorption of latent heat. The cooling effect of the liquid is either used for the air-conditioning or compensated with calories from outside. The system can be regenerated by heating up the reactor with resistors supplied by the battery charging station ; the above reaction is inverted, and the vapor V condensates in the evaporator with evacuation of heat which can be used in winter for pre-heating.

Fig 7. Reversible air-conditioning

3. Separate energy storage systems

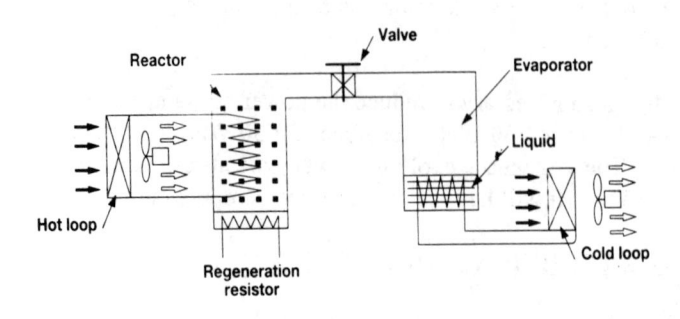

Fig.8 Sorption Systems

The table II gives some values of the ratio energy/weight for different systems compared to batteries of different types.

Table II - Energy storage ratio
for some different systems

Type of storage	Energy/weight $10^3 \times J/kg$
Pb Battery	144
Cd/Ni Battery	198
Na/S Battery	396
Latent heat storage	216
Sorption storage	
Water / Zeolithe : Heating	540
Air-conditioning	288
Metallic Chloride / NH_3	
Heating	648
Air-conditioning	360

The STELF system, using the metallic chloride/graphite matrix as solid reactive with NH3 is very interesting due to a very good ratio and a high thermal power.

4. Recovery of heat rejections from the traction chain (when they are available)

These rejections can be significant, and are interesting to be recovered, either directly upstream a PTC resistor, or with a heat pump. It is however necessary to take into account the fact that, in summer, these rejections must be dissipated ; their management needs an electronic control.

On a hybrid vehicle, it is also possible to use the heat rejection from the engine in the same way.

CONCLUSION

Technical solutions exist to solve the problem of climatic comfort in electric and hybrid vehicles. However, it is necessary to analyse the specifications, accepting eventually lower performances at extreme outside temperatures.

Bibliography

[1] *Improved vehicle heating using PTC ceramic.* Automotive Engineering, March 1990, pp.55-59.

[2] DUMINIL (M) ; *Systèmes à absorption, à adsorption et thermochimiques en vue de la climatisation.* Revue Générale du Froid, Numéro spécial Juin 1991, pp. 74-75.

[3] Nissan FEV ; Japan Autotech. Report, Vol 131, October 20, 1991, p29.

[4] *Latent heat storage.* Automotive Engineering February 1992, pp. 58-60.

[5] DICKMANN (J.), MALLORY (D.) ; *Climate Control for Electrical Vehicles.* SAE Paper 910250.

[6] DICKMANN (J.), MALLORY (D.) ; *Variable Speed Compressor HFC-134a based air conditioning system for electric vehicles.* SAE Paper 920444.

[7] BURK(R), KRAUSS (H.J), LOEHLE (M) ; *Intergrales Klymasystem für Electroautomobile.* ATZ Automobiltechnische Zeitschrift. Vol. 94 (1992).

940296

Proton Exchange Membrane Fuel Cell Characterization for Electric Vehicle Applications

D. H. Swan, B. E. Dickinson, and M. P. Arikara
University of California, Davis

ABSTRACT

This paper presents experimental data and an analysis of a proton exchange membrane fuel cell system for electric vehicle applications. The dependence of the fuel cell system's performance on air stoichiometry, operating temperature, and reactant gas pressure was assessed in terms of the fuel cell's polarity and power density-efficiency graphs. All the experiments were performed by loading the fuel cell with resistive heater coils which could be controlled to provide a constant current or constant power load. System parasitic power requirements and individual cell voltage distribution were also determined as a function of the electrical load. It was found that the fuel cell's performance improved with increases in temperature, pressure and stoichiometry within the range in which the fuel cell was operational. Cell voltage imbalances increased with increases in current output. The effect of such an imbalance is, however, not detrimental to the fuel cell system, as it is in the case of a battery.

INTRODUCTION

An electrochemical fuel cell is a device that converts chemical energy to direct current electrical energy. By converting an on-board fuel to electricity it could be effectively used to power an electric vehicle. As such, a fuel cell is an energy conversion device like an internal combustion engine. This is in contrast to energy storage systems such as batteries, flywheels and ultra capacitors. Further many of a fuel cells operating characteristics are closer to that of an engine than a storage battery. A fuel cell system operation involves startup, fuel and air delivery control as a function of load, and removal of heat and by products of the reaction. The fuel cell, in other words, is an electrochemical engine. While electrochemistry describes the principle of operation of a fuel cell, the engineering challenge of balancing the many variables over a wide variety of operating conditions remains. The fuel cell system consists of a complex group of support systems that must operate in balance for efficient performance.

Different types of fuel cells are conveniently classified by the type of electrolyte they use. Electrolytes that are presently being considered include the proton exchange membrane (a solid polymer material), phosphoric acid (a liquid), alkaline (a liquid), molten carbonate (a liquid) and solid oxide (a ceramic). The choice of electrolyte directly affects a fuel cell's operating characteristics; for example, phosphoric acid is a poor ion conductor at room temperature. As a result the phosphoric acid fuel cell must be heated to 150 to 200°C before it can be used. Today many researchers believe that the proton exchange membrane (PEM) provides the best characteristics for transportation applications.

The data and analysis presented in this paper is for a fuel cell system manufactured by Ballard Power Systems of Vancouver, Canada. The fuel cell system consists of: a 35 cell series connected stack; gas, water and thermal management subsystems; and controls and monitors all assembled in a single enclosure. The area of each cell was 232 cm^2 and the fuel cell stack itself had a maximum gross power output greater than 3000 Watts operating on hydrogen and air. The system was modified by the authors to be able to independently control air stoichiometry, air/hydrogen pressure and stack exit air temperature. Previous papers that have presented experimental data on similar Ballard fuel cells systems are references[1] and [2]

The paper is organized into sections in the following order; Fuel Cell Operating Principle, Experimental Apparatus, Experimental Results, Results Analysis and Conclusions. The section on fuel cell operating principle is intended to give a brief overview of how a fuel cell works and its operating characteristics. The experimental apparatus section briefly describes the fuel cell system used in the experiments and the associated instrumentation. The experimental results present a series of polarity plots (voltage - current relationship) under a variety of operating conditions. The results analysis section presents the results in terms of power density-efficiency plots implicitly demonstrating the operating characteristics of the fuel cell system for electric vehicle applications.

FUEL CELL OPERATING PRINCIPLE

All energy-producing oxidation reactions are fundamentally the same and involve the release of chemical energy through the transfer of electrons. During combustion of hydrogen and oxygen there is an immediate transfer of electrons, heat is released and water is formed.

In a fuel cell the hydrogen and oxygen do not immediately come together but are separated by an electrolyte. First the electrons are separated from the hydrogen molecule by a catalyst (oxidation) creating a hydrogen ion (no electrons). The ion then passes through the electrolyte to the oxygen side. The electrons cannot pass through the electrolyte and are forced to take an external electrical circuit which leads to the oxygen side. The electrons can provide useful work as they pass through the external circuit. When the electrons reach the oxygen side they combine with the hydrogen ion and oxygen creating water. By forcing the electrons to take an external path, a low temperature direct energy conversion is achieved as shown in Figure 1.

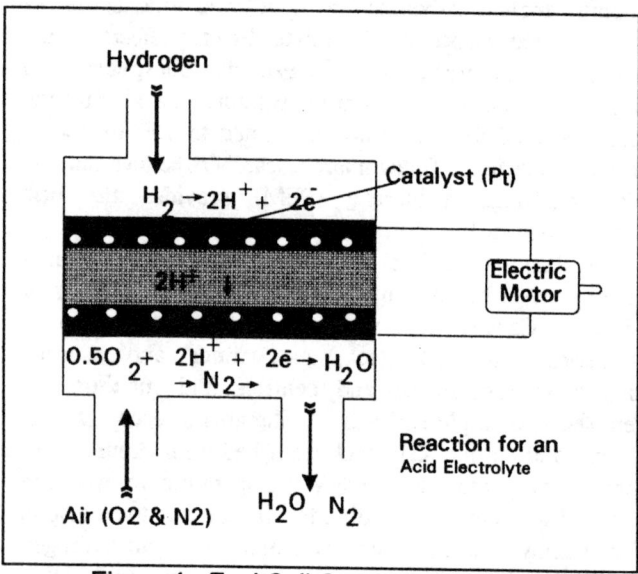

Figure 1. Fuel Cell Operation Schematic

The theoretical efficiency for the conversion of chemical energy into electrical energy in a hydrogen-oxygen fuel cell depends on the free energy of formation for the reaction (Gibbs Function). The free energy of formation is equal to the difference of the heating value of the fuel and its entropy at the temperature and pressure of conversion. This is described in equation 1.

$$\Delta G = \Delta H - T \times \Delta S \quad (1)$$

Where

ΔH Free energy of Formation

T Absolute Temeperature

ΔS Entropy

Typically theoretical chemical to electrical conversion efficiencies are in the range of 83% for higher heating value and 94% for the lower heating value of hydrogen. Efficiencies of practical fuel cells using pure hydrogen and air range from 40% to 75% based on lower heating value of hydrogen.

Assuming a near perfect coulombic efficiency (all electrons are forced to take the external circuit) a theoretical operational voltage can be calculated. This voltage is calculated by considering Faraday's constant (26.8 Amp hours = 1 mole of electrons) and the energy value of the fuel. Table 1, provides the hydrogen/oxygen-water reaction enthalpy (heating value), the free energy of formation[3] and the resultant theoretical cell voltages.

Table 1   Hydrogen Thermodynamic Properties

Heating Value	ΔH Enthalpy of Formation kJ/Mole	Gibbs Function $\Delta G = \Delta H - T \Delta S$ kJ/Mole At 25°C, 1 bar	Voltage Based on Enthalpy	Cell Voltage Based on Gibbs
Higher	-285.9	-237.2	1.48	1.23
Lower	-241.8	-228.6	1.25	1.18

Like a storage battery, when the fuel cell is under electrical load the voltage falls with maximum power generally being produced between 0.5 and 0.6 volts per cell. The voltage drop as a function of current is due to internal resistance (electronic and ionic), electrode kinetics (particularly on the air electrode), reactant gas flow limitations and product water flooding of reaction sites. To make a useful voltage, multiple cells are connected in electrical series, referred to as a stack as shown in Figure 2. Manifolds deliver reactant gases to the reaction sites. The fuel cell stack and all necessary auxiliaries are referred to as a fuel cell system.

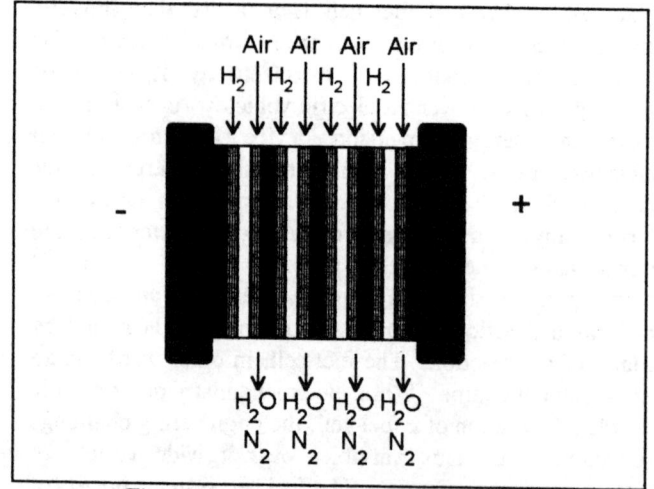

Figure 2. Fuel Cell Stack Schematic

The fuel cell stack design must allow for heat exchange and humidification of incoming reactant gases, thermal management, product water management, exhaust gases, and electrical management.

FUEL CELL SYSTEM DESCRIPTION - Given below are some of the specifications for the fuel cell power system which among others includes a 35 cell stack with a humidification section, temperature management, reaction product management and electrical control.

Table 2 Fuel Cell System Specifications.

Electrolyte	Nafion-117
# of Active Cells	35
Active Area/cell	232 cm^2
Total active area	8120 cm^2
Active Cell Thickness	0.5 cm
# of Cooling Cells	19
Active Plate Thickness	0.5 cm
# of Humidification Cells	14
System dimensions LxWxH	104 x 80 x 37 cm
System volume	307.8 liters
Stack dimensions LxWxH	38 x 21 x 21 cm*
Stack volume	16.76 liters*
Weight of the stack	43 kg
Weight of the system	50 kg
Support System Power **	350 Watts

*Includes active stack with humidification section and cooling cells
**water pump, hydrogen recirculation pump, fans, solenoid valve control and monitoring system

DATA ACQUISITION

One of the important tasks in the experiments conducted was to monitor and control the operating conditions of the fuel cell. The data acquisition and control system used signal conditioning modules which were capable of either data acquisition or control based on commands received from an IBM compatible personal computer. The modules communicated with the computer through an RS-232 communications port. Each module was isolated to 1500 volts and includes a filter to reduce noise. The control system involved both digital and analog input and output.

The load for the fuel cell was provided by resistive heater coils. The resistance provided to the fuel cell was controlled by a pulse width modulator (PWM). The PWM in turn was controlled by the analog output of the control system. The resistance could be controlled to provide a constant power or constant current load.

The variables measured by the control system included, stack current and voltage, support system current, load current, process air exit temperature, hydrogen inlet pressure, air inlet and outlet pressure. The digital signals monitored included the status of the load relay, hydrogen vent, oxidant vent and the water drain.

On the control side the load applied to the fuel cell was controlled by an analog output module. The digital output controlled the heat exchanger operation in order to maintain the temperature of the fuel cell at the required levels.

In order to maintain the stoichiometry of the air provided for the reaction a needle valve provided on the fuel cell was controlled. The flow rate was monitored by a flow meter installed in the air delivery line. Pressure control was established by using single stage pressure regulators.

Hydrogen for the fuel cell stack was provided from industrial hydrogen cylinders. The air system included a two stage compressor along with a chiller dryer and a system of air filters.

EXPERIMENTAL RESULTS

The experiment was organized into exploring the effects of three independent variables on performance: air stoichiometry, air pressure and exit air temperature. Air stoichiometry is the ratio of the amount of air provided to the fuel cell to that which is necessary to react with the hydrogen fuel. Other words used to describe this ratio include Lambda and excess air factor. In a typical internal combustion engine the value is very close to 1. Due to the operating nature of a fuel cell the stoichiometry of the air must be greater than one (typically 1.5 to 4). A lower value of stoichiometry reduces performance due to the lack of oxygen at the reaction sites. A high value of stoichiometry results in poor humidity control and excess compression energy.

The following test matrix illustrates the 12 different operating conditions over which the fuel cell system was characterized. The stoichiometric accuracy was held between +/- 0.1, the pressure accuracy was held between +/- 0.05 bar and the temperature accuracy between +/- 1.5°C. Under each operating condition the stack output current was varied between a minimum and maximum amperage at 10 amp intervals. The minimum current condition (20 amps) was limited by the support system power requirements and a small load on the electric load bank. The maximum current condition was limited by a minimum stack voltage of 19 (0.54 volts/cell). The minimum stack voltage was a limit set by Ballard power systems and was within a safe operating range for continuos operation of the system.

Table 3 Test Matrix For Fuel Cell Testing

Stoichiometry	Hydrogen and Air Pressure in Bar	Stack Exit Air Temperature °C
1.5	2	50, 60, 70
	3	50, 60, 70
2.0	2	50, 60, 70
	3	50, 60, 70

The results of this test matrix are organized into two polarity plots (Stoich.=1.5 and Stoich.=2.0), and an individual cell voltage characterization

POLARITY PLOTS - The polarity data resulting from the Table 2 test matrix is presented in Figures 3 and 4. These plots include the measured full stack data (volts and amps) and normalized data (volts/cell and milli amps/cm^2 of active area) The normalized data represents an average for the stack; actual individual cell data is presented in the following sub section.

Figure 3 presents the results for the six cases for which the stoichiometry was equal to 1.5. Figure 4 presents the same results with the stoichometry equal to 2.0. The current range over which the two plots are described is the normal operating range of the fuel cell system. (Open circuit voltage, although not shown, was found to be approximately 1 volt per cell.).

Both, Figures 3 and 4, indicate a near linear relationship between voltage and current. The higher stoichiometry shown in Figure 4 results in a higher voltage at any given current compared to Figure 3. The difference in voltaic performance between the two stoich. rates is most pronounced at the high current levels. With a Stoich of 1.5 the fuel cell system could only deliver 130 amps before the low voltage limit was reached. In comparison, at a stoich rate of 2.0 the fuel cell could deliver approximately 160 amps. At a reference performance condition of 100 amps (normal operating condition) the average voltage increase with increase in stoich rate from 1.5 to 2 was 1.35 volts or 6.0 %. At lower current rates the percentage difference is smaller than at the higher rates. This seemingly small increase in voltage at a given current has the two-fold effect of increasing power and operating efficiency. In the following section; 'Results Analysis', it will be shown that even a relatively small increase in voltage has a significant impact on performance for an electric vehicle.

As a function of pressure, voltaic performance also increased. Once again utilizing a reference performance condition of 100 amps the typical voltage increase going from 2 to 3 bar was between 0.5 and 1.2 volts (2 to 5%). Voltaic performance also increased as a function of temperature. Once again utilizing a reference performance condition of 100 amps the typical voltage increase observed for an increase in temperature from 50 to 60 °C was 0.5 to 1.0 volts (2 to 4%), for an increase from 60 to 70 °C the corresponding voltage increase was 0.3 to 0.7 volts (1 to 3 %). The benefit of going from 60 to 70°C was less than going from 50 to 60°C There will be an optimum operating temperature for the fuel cell beyond which the voltaic performance will decline. The experimental data indicates that this value is probably greater than 70°C.

For both the pressure and temperature changes the percentage change in voltaic performance is greater at high current rates. Once again this seemingly small increase in voltage at a given current has the two-fold effect of increasing power and operating efficiency. Thus a relatively small increase in voltage has a significant impact in performance for an electric vehicle.

INDIVIDUAL CELL PERFORMANCE - The voltage performance of individual cells was measured at a exit air temperature of 60°C and a pressure of 3 bar. Five different stack current rates were used (20, 40, 70, 100, 130 amps) and the respective cell voltages are presented in Figure 5. The measurements were manually made utilizing a digital volt meter and pointed probes directly on to the active cell plate edge.

Figure 5 shows that there is a difference in voltaic performance for each cell at a given current rate. The differences in performance are consistent over the operating current range, a cell that shows a lower voltage at a low current density continues to have a lower voltage at higher current densities. The lower voltage could be the result of the difference in the electrode membrane assembly, distribution of reactant gases, internal cell resistance or a persistent flooding effectively reducing the available reactive area.

Although not shown, individual cell performance has been measured several times over the 6 months of preparation to conduct the experiments described in this paper. Consistently cells 6 and 25 have had a lower voltaic performance than the other cells (0.05 volts lower than average for 100 amps). As a result the authors feel that difference is inherent to the particular cells and not a transient phenomena due to a flooding condition or other temporary phenomena. It is interesting to note that cells 7, 24 and 26 perform above average, these high performance cells are adjacent to the lowest performing cell and may indicate that the problem is one of flow distribution.

The following Table 4 summarizes the results of the individual cell measurements. Although Figure 5 shows a consistent pattern of cells 6 and 25 having lower than average voltage the overall stack performance is very good. The maximum voltage spread is the result of cells 6 and 25 while the standard deviation is very small due to the generally consistent performance of the stack.

Table 4 Individual Cell Voltage Characteristics

Stack Current	20	40	70	100	130
Stack Voltage	30.25	28.4	26.5	24.4	22.4
Current Density mA/cm^2	86	172	302	431	560
Average Cell Voltage	0.864	0.812	0.758	0.698	0.640
Maximum Voltage Spread	0.03	0.042	0.063	0.086	0.115
Cell Voltage Standard Deviation Volts	.006	.008	.0115	.017	.025

SUPPORT SYSTEMS ELECTRICAL LOAD - The fuel cell system has support components that are electrically operated and controlled. Support system power is drawn from the fuel cell stack through a voltage regulated power supply (constant 26 volt output). The support components include a combined cooling and reactant gas humidification system that recovers product water from the stack exit air, a hydrogen recirculation system, and a control system that monitors and controls the start up, operation and shut down of the fuel cell system.

The combined cooling and humidification system utilizes a circulation water pump that provides water to cooling plates followed by the reactant gas humidification section. Based on stack exit air temperature the water leaving the humidification section is routed through a water to air heat exchanger. The hydrogen recirculation system maintains a constant recirculation rate through the stack and a liquid water knock out drum. This circulation combined with periodic purges prevents flooding conditions and build up of inert gas on the hydrogen side of the fuel cell stack.

Table 5. gives the support system load demand during operation of the fuel cell. The support system power requirement does not include air compression which is independent of the fuel cell stack output. As a result the support system power (a parasitic loss) varies in percentage of stack power from 100% (idle condition) to 10% (full power condition. The difference in support system power (250 to 400 Watts) results from the cooling of the circulating water. When the water is circulated through the water to air heat exchanger the additional power is drawn by ventilation fans.

Table 5   Support Systems Loads

Component	Amps
Water Pump	3-4
Hydrogen Circulation pump	1
Ventilation Fans	2
Controller, batteries and solenoids	3-6
Total Support System	10 to 15 (250 to 400 Watts)

RESULTS ANALYSIS

The previous sections described fuel cell polarity plots, which is a standard method to describe fuel cell performance. However, the interests of an automotive engineer in applying a fuel cell to an electric vehicle is better described in terms of power and efficiency. The specific power available determines the size of the fuel cell system and to some extent its cost. The conversion efficiency affects the sizing of the fuel storage, resultant range and refueling times.

To accommodate this need for a power and efficiency relationship the polarity data can be reinterpreted. The specific power of the fuel cell is simply the ratio of the product of the stack voltage and the stack current to the total available active electrolyte area.

$$\textbf{Specific Power} = \frac{V \times I}{\textbf{Total Active Area}} \times 1000 \quad \left(\frac{mWatts}{cm^2}\right) \quad (2)$$

Where
V Stack Voltage
I Stack Current

To calculate the fuel cell stack gross conversion efficiency it is assumed that the system is operating at a perfect coulombic efficiency (all available electrons from the hydrogen fuel were forced to take the external circuit providing useful work). This means that all inefficiencies are manifested in a voltaic loss. Equation 3 presents the method used to calculate fuel cell stack conversion efficiency. This approach to calculating efficiency is generally considered valid for the proton exchange membrane fuel cell, however measurements will be made in the future using this same fuel cell system to quantify this assumption. See the earlier section, 'Fuel Cell Operating Principle', for further details. Note that no allowance is made for the support system power requirements or air compression. As a result the net efficiency of conversion (ratio of delivered electrical energy to chemical energy) is less than the fuel cell stack conversion efficiency.

$$\textbf{Fuel Cell Stack Conversion Eff.} = \frac{V}{1.25 \times 35} \times 100 \% \quad (3)$$

Where
V   Stack Voltage
1.25   Theoretical Cell Voltage
       Based on Enthalpy of Formation   (Lower)
35   Number of Cells in Stack

Utilizing equations 2 and 3 the data presented in Figures 4 and 5 was recalculated and presented in Figures 6 through 9. Figures 6 through 9 present the fuel cell stack specific power as a function of fuel cell stack conversion efficiency based on the lower heating value of hydrogen. In these figures a higher electrode power density translates to a smaller electrode area to achieve a given power level. A high conversion efficiency translates to a smaller fuel storage and refueling time for a given vehicle range and performance. A second order affect is that a higher efficiency translates to smaller heat generation within the fuel cell system thus a smaller heat exchanger system and other support components.

Figures 6 through 9 are organized as a function of stoichometry and operating pressure. Figures 6 and 7 present specific power data for a stoich of 1.5 and pressures of 2 and 3 bar respectively. Each figure shows the results for the three different temperatures in the test matrix (50 60 and 70 °C).

Consider Figure 6, the specific power-efficiency curves indicate a trade-off between power and efficiency. The higher the specific power density, the lower is the conversion efficiency. The relationship between specific power and efficiency appears linear for high conversion efficiencies, 70% to approximately 55%. At this point the power increase for the drop in efficiency becomes less and the curve starts to flatten. Although the curve does not distinctly show a specific power peak the data indicates that this peak is somewhere around 45% energy conversion efficiency, this corresponds to an average cell voltage of approximately 0.55 (just slightly greater than one half the open circuit voltage). Increasing the current beyond the peak power point results in a further decline in specific power and conversion efficiency. At no time would it be advantageous to operate the fuel cell system beyond the peak power point.

The following analysis utilizes three different specific power conditions to illustrate the influence of stoichometry, pressure and temperature. The three specific power conditions selected are 125 mW/cm^2, 275 mW/cm^2, and 325 mW/cm^2. The low power condition (125) relates to a possible cruise condition for the electric vehicle, the high power condition (275) relates to acceleration or hill climbing. The highest specific power 325 was not attainable by the system at low stoichometry and pressure and is used to illustrate the enhanced capability by changing operating conditions. Table 6 relates the energy conversion efficiency to the three selected specific power conditions.

First comparing the affect of stoichometry, the efficiencies of operation in Table 6 immediately shows that at low power conditions stoichiometry has little influence on efficiency of operation resulting only in 1 to 2 percentage points increase. At the higher power condition the spread increases to 4 to 5 percentage points or in other words a 10% change in stack fuel consumption for the same power level. At the highest power condition the increase in stoichiometry is necessary for the stack to reach the 325 mW/cm^2 specific power level The low stoichiometry simply cannot reach the high level having peaked at approximately 290 mW/cm^2. Increased stoichiometry is most beneficial at high power conditions.

Table 6 Fuel Cell Stack Conversion Efficiency at Three Specific Power Points

Operating Condition	Speci. Power 125 mW/cm^2	Speci. Power 275 mW/cm^2	Speci. Power 325 mW/cm^2
S=1.5, T=60C, P =2 Bar	64%	51%	not possible
S=1.5, T=60C, P =3 Bar	64%	54%	46%
S=2.0, T=60C, P =2 Bar	65%	56%	50%
S=2.0, T=60C, P =3 Bar	66%	58%	53%

Comparing the influence of an air pressure change from 2 bar to 3 bar there is almost no difference for the low power case. Under high power conditions the change in pressure increases the efficiency by 2 to 3 percentage points or a difference of 6% in stack fuel consumption for the same power level. For the highest power condition a pressure change from 2 to 3 bar allows the lower stoichometry of 1.5 to reach high power of 325 mW/cm^2, however the lower stoichiometry has a energy conversion efficiency 7% points lower or a stack fuel consumption of 15% higher. Like stoichiometry the benefit of pressure is most pronounced at high power conditions. Both increased pressure and stoichiometry increase the compression power needed to operate under these conditions. As a result, choosing the stoichiometry and operating pressure is not a simple straight forward procedure. The affect of air compression power will be examined in a following section.

Table 7 compares the influence of temperature on performance for a stoichiometry of 2, the two different pressure cases are presented for comparison. Once again at a low specific power there is little difference in efficiency. At the high power level the efficiency spread resulting from an increase temperature is between 3 and 5% points. At the highest power level the low temperature low pressure case can not achieve 325 mW/cm^2. By increasing the temperature to 70°C the highest power condition was possible at a conversion efficiency of 51%. The effect of temperature rise on the 3 bar pressure case was to increase efficiency by 3 % points, or a difference in stack fuel consumption of 6%. There will be an optimum operating temperature for the fuel cell beyond which the conversion efficiency for a given specific power will decline. The experimental data indicates that this value is probably greater than 70°C. The high operating temperature reduces the sizing of the fuel cell cooling system by a. Further there is no additional support energy required to maintain the fuel operating temperature at a high value.

Table 7 Fuel Cell Stack Conversion Efficiency at Three Specific Power Points

Operating Condition	Speci. Power 125 mW/cm^2	Speci. Power 275 mW/cm^2	Speci. Power 325 mW/cm^2
S=2.0, T=50C, P =2 Bar	64%	52%	not possible
S=2.0, T=70C, P =2 Bar	65%	57%	51%
S=2.0, T=50C, P =3 Bar	66%	56%	52%
S=2.0, T=70C, P =3 Bar	67%	59%	55%

This analysis also indicates the clear advantage that a fuel cell has at part load condition. Unlike an internal combustion engine the energy conversion efficiency increases substantially as the load is reduced (typically 15 percentage points or a fuel reduction of 25% for a kWh produced). Not evident in this analysis is that highly dynamic operation does not degrade the efficiency. Operating on hydrogen the fuel cell responds instantaneously to the new operating load with no loss in efficiency for the change in power condition[4]. A difficult that does arise under dynamic conditions is the need to control the air flow at prescribed stoichiometric rates.

This combination of high part load efficiency and no degradation due to dynamic operation gives the fuel cell a clear advantage over the internal combustion engine. This advantage is particularly important for a city driving cycle.

AIR COMPRESSION ENERGY - The support system for a fuel cell power plant include controls, cooling fans, recirculation pumps (water and hydrogen) and compressed process air. The difference between fuel cell stack power and fuel cell system power is due to the parasitic losses associated with the support system components. Of these components and the energy required to operate them the compressed air energy requirement is by far the largest.

Pressurization of air improves the performance of the fuel cell when considered as an individual entity. However, the energy for compression on an electric vehicle must be supplied by the fuel cell system, and thus net power output is less than the gross stack performance. The air compression process can be adiabatic or isothermal and part of the compression energy may be recovered from the exiting air by an expander such as a turbine. To determine the influence the energy of compression would have on the fuel cell performance the following analysis assumes that the compression is performed by an ideal adiabatic compressor with no energy recovery upon expansion. The affect of the air compressor can be related to a simple loss in net cell voltage as a function of pressure ratio and stoichiometry. A full derivation may be found in reference[5] a brief explanation is provide in the following paragraph.

Air compression power is a function of inlet to outlet pressure ratio and the quantity of air compressed. The relationship of compressor power to quantity of air compressed is linear. For a constant stoichiometric the quantity of air needed is directly related to the current being produced. Double the current, double the air is needed to maintain the stoichiometry, double the compressor power is needed to maintain the flow. Consider an ideal compressor in series with the electrical load. When the fuel cell current is increased the compressor power must increase in direct proportion. Because the current has gone up the voltage drop across the compressor remains the same. The voltage available to the electrical load (net voltage) is simply the cell voltage minus the effective compressor voltage that is a constant for a give stoichiometry and pressure ratio. Thus the effect of air compression can be simply presented effectively as a reduction in voltage. Utilizing equation 3, this effective compressor voltage loss can be used to determine the impact air compression has on efficiency. The following equation to calculate effective compressor voltage was first derived in reference[6]

$$V_c = \frac{1.287}{3600} \times \text{\# of Stoich.} \times C_p T_1 \left(\left(\frac{P_2}{P_1}\right)^{\frac{k-1}{k}} - 1 \right) \quad (4)$$

Where
V_c Effective Adiabatic Compressor Voltage loss per Cell
C_p Specific Heat (Air 1.004 J/(g °K))
T_1 Compressor Inlet Air Temperature °K
P_1 Compressor Inlet Air Pressure
P_2 Compressor Outlet Air Pressure
k Specific Heat Ratio (Air 1.4)

Utilizing equation 4 the effective compressor voltage loss for the 35 cell stack data presented in this paper is tabulated in Table 8. The voltage loss has been interpreted as an efficiency reduction and is presented in brackets.

Table 8 Effective Adiabatic Compression Voltage Loss and Resultant Efficiency Loss

Stoich	Press = 2 Bar	Press = 3 Bar
1.5	1.23 volts (-2.8%)	2.07 volts (-4.7%)
2	1.64 volts (-3.7%)	2.77 volts (-6.3%)

Table 8 indicates that the affect of the energy of air compression is to reduce operating efficiency by 2.8 to 6.3 percentage points. Considering that air must be forced through the fuel cell stack the 2 Bar case will be considered as a base lines. Increasing the pressure from 2 to 3 bar increase the effective degradation in efficiency by 1.9 to 2.6 percentage points.

In the previous subsection the influence of a pressure change from 2 to 3 bar on efficiency at low specific power conditions was found to be 1 percentage point or less, thus the net effect of compression would be negative. At high power conditions the affect of pressure change from 2 to 3 bar is to increase conversion efficiency by 3 percentage points. This gain is almost entirely nullified by the effective compressor efficiency loss.

Utilizing the data for a stoichiometry of 2 at 60°C the performance of pressure equal to 2 bar and 3 bar is compared in Figure 10. The fuel cell stack specific power for the 3 bar case is higher then the 2 bar case. However when the effective compressor voltage loss is considered (Table 8 values) the performance of the 2 bar case exceeds the 3 bar case. The 3 bar case has a higher peak power value but it is notably lower efficiency.

Air compression decreases the conversion efficiency of fuel cell system. It also impacts the peak specific power available. A high stack performance from increasing the operating pressure may not be enough to over come the additional power requirement of the compressor.

CONCLUSIONS

The 3 kW fuel cell system described in this paper is an impressive performer with approximately 40 hours of operation in preparing and obtaining the presented data. Measured energy conversion efficiencies while operating on hydrogen and air ranged between 45% and 70%. Under part load conditions (1 to 2 kW the efficiency typically ranged between 55 and 65% , peak power output (3 kW) typically occurring at 45%. The trade off between power and efficiency will influence the sizing of the fuel cell system. For a given vehicle application a larger more expensive fuel cell will operate at a lower average power and thus have a better fuel economy.

The result of increasing the fuel cell operating pressure and stoichiometry is to increase its performance. The influence of these two parameters was found to be dependent on the fuel cell electrical load. At low load conditions increasing stoichiometry or pressure had little influence. At high load conditions increasing the stoichiometry from 1.5 to 2 improved performance by as much as 7 percentage points. This would effectively reduce stack fuel consumption by 15% for the same amount of energy conversion. Increasing the operating pressure from 2 to 3 bar would improve stack performance as much as 3 percentage points or an effect fuel consumption drop of 7%.

The benefits of increased stoichiometry and pressure must be balanced against the power needed to compress the additional air to a higher pressure. A brief analysis of the calculated adiabatic compression power with no pressure recovery was presented. The analysis found that based on the experimental data there was no net advantage in increasing the operational air pressure from 2 to 3 bar. This calculated result indicates the trade off between stoichometry, pressure and the energy of compression is not a simple one. Other support components that must be powered by the fuel cell include controls, a water pump and a hydrogen recirculation pump. These components required a near constant 250 to 350 Watts (10% of maximum fuel cell stack power) and were independent of the fuel cell system power output. Both pumps were operated at constant rates and sized for maximum operating conditions. In an automotive design they will be variable and thus reduce the percentage power requirements of the pumps significantly.

The affect of increasing the operational temperature (stack exit air temperature) from 50 to 60 to 70 °C was an improvement in performance. There is an optimum operating temperature for the fuel cell stack beyond which the performance will decline. The experimental data indicates that this value is probably greater than 70°C. High operating temperature reduces the sizing of the fuel cell cooling system and is considered an advantage.

Individual cell voltaic performance was measured and found to have slight variations between cells. Two cells in particular had lower voltages (0.1 volts below average at a current of 130 amps). Adjacent to these cells were cell performing above average, indicating possible reactant gas flow distribution problem. Unlike a storage battery the lower voltaic performance is not a serious problem.

The fuel cell power system is a strong candidate to meet the needs of California's ZEV mandates. Utilizing on-board hydrogen the only byproduct of operation is water. The high conversion efficiency of a fuel cell overcomes many of the storage and cost problems associated with hydrogen. The promise of fuel cell technology is a ZEV with the performance, range and rapid refueling capability of conventional vehicles.

ACKNOWLEDGMENTS

The authors wish to thank Ballard Power Systems and California Department of Transportation (New Technologies Division) for the support they have provided in setting up the fuel cell laboratory at the Institute of Transportation Studies, Davis. We also wish to thank the University of California Transportation Center for financially supporting the Graduate Students involved in this project. Last but not the least we wish to thank Mr. Gonzalo Gomez and Mr. Manohar Prabhu for their assistance in setting up the data acquisition and control system and helping us in obtaining the experimental data.

REFERENCES

[1] Prater K. B., "Solid Polymer Fuel Cell Developments at Ballard", Procedings of the 2nd Grove Symposium, Editors A.J. Appleby and D.G. Lovering, 1991 p 189 to 201.

[2] Olivera C.T., A. Anantaraman and W.A. Adams, "Performance Evaluation of a H2/Air PEM-FC System under Variable Load", Precedings of the 1992 Fuel Cell Seminar, Tucson, Arizons, December 1992 p 451 to 454

[3] Masterton W.L., E.J Slowinski, "Chemical Principles", W.B. Saunders Co. Philadelphia PA, 1973.

[4] Dickinson B.E., T. Lalk, D. G. Hervey, "Characterization of a Fuel Cell/Battery Hybrid System for Electric Vehicle Applications", SAE Paper 931818, published in Special Publication 984, 1993.

[5] Swan D.H , A.J. Appleby, "Fuel Cells for Electric Vehicles, Knowledge Gaps and Development Priortities", Proceedings of The Urban Electric Vehicle, Stockholm Sweden, pp 457-468, May 1992.

[6] Swan D.H., O.A. Velev, I.J. Kakwan, A.C. Ferreria, S. Srinivasan, A.J. Appleby, "The Proton Exchange Membrane Fuel Cell - A Strong Candidate as A Power Source for Electric Vehicles", Hydrogen 91 Technical Proceedings, International Association for Hydrogen Energy, 1991.

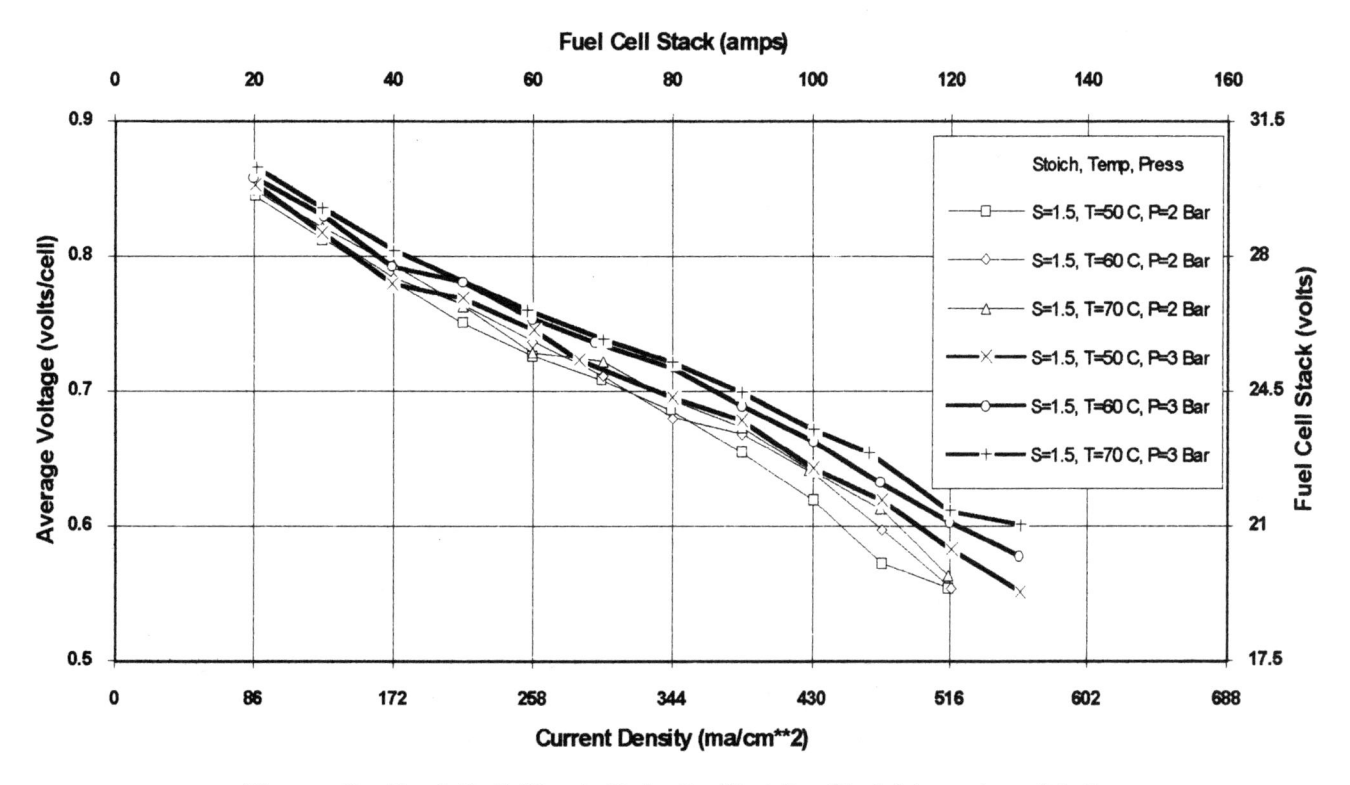

Figure 3. Fuel Cell Stack Polarity Plot for Stoichiometry of 1.5

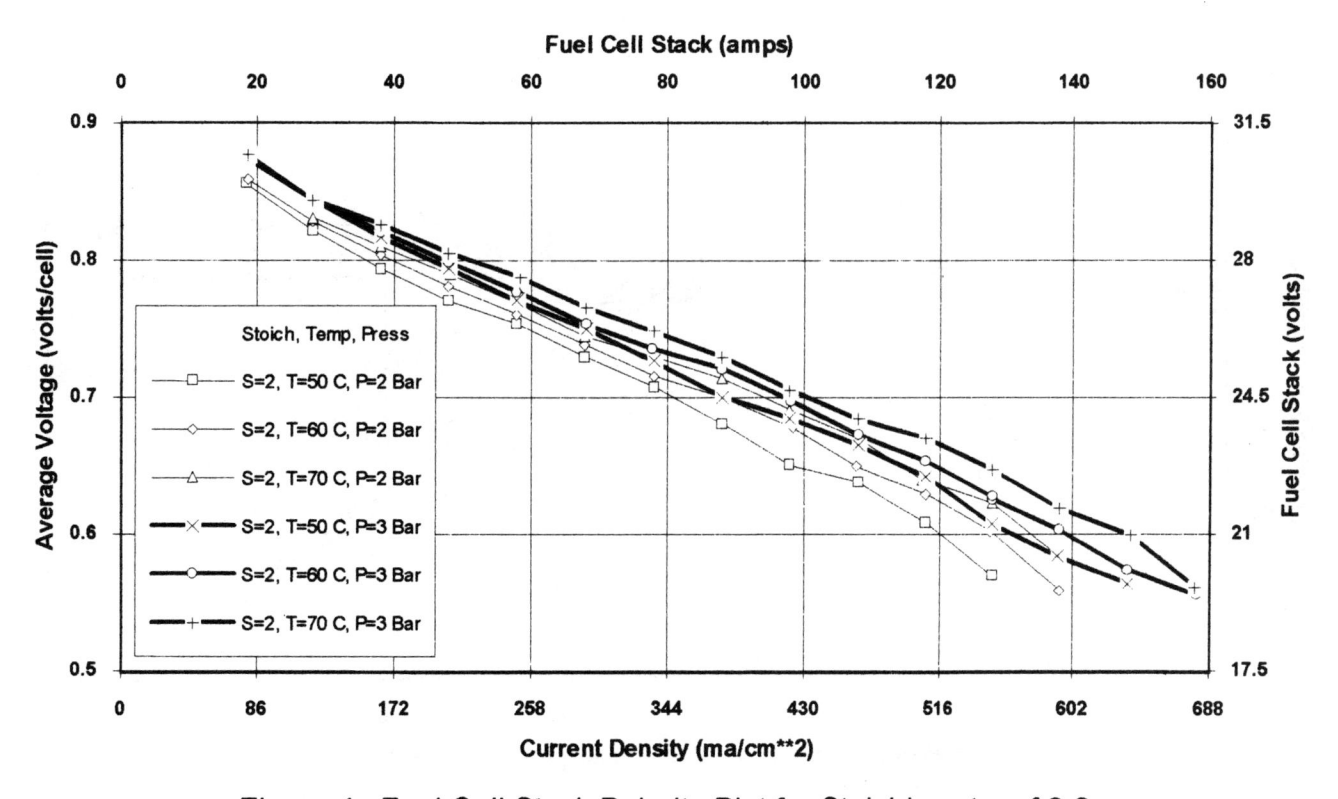

Figure 4. Fuel Cell Stack Polarity Plot for Stoichiometry of 2.0

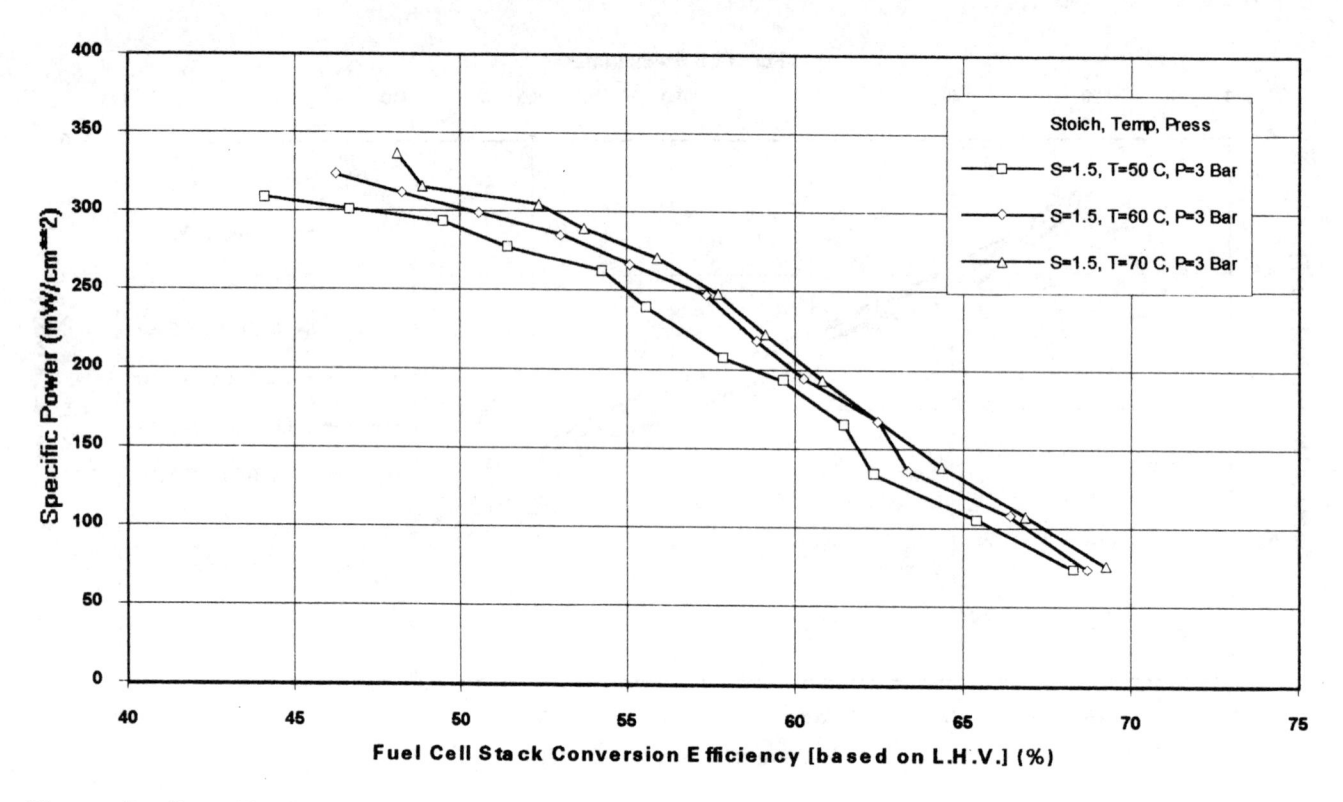

Figure 7. Specific Power as a Function of Temperature and Efficiency for Stoichiometry of 1.5, Pressure = 3 Bar

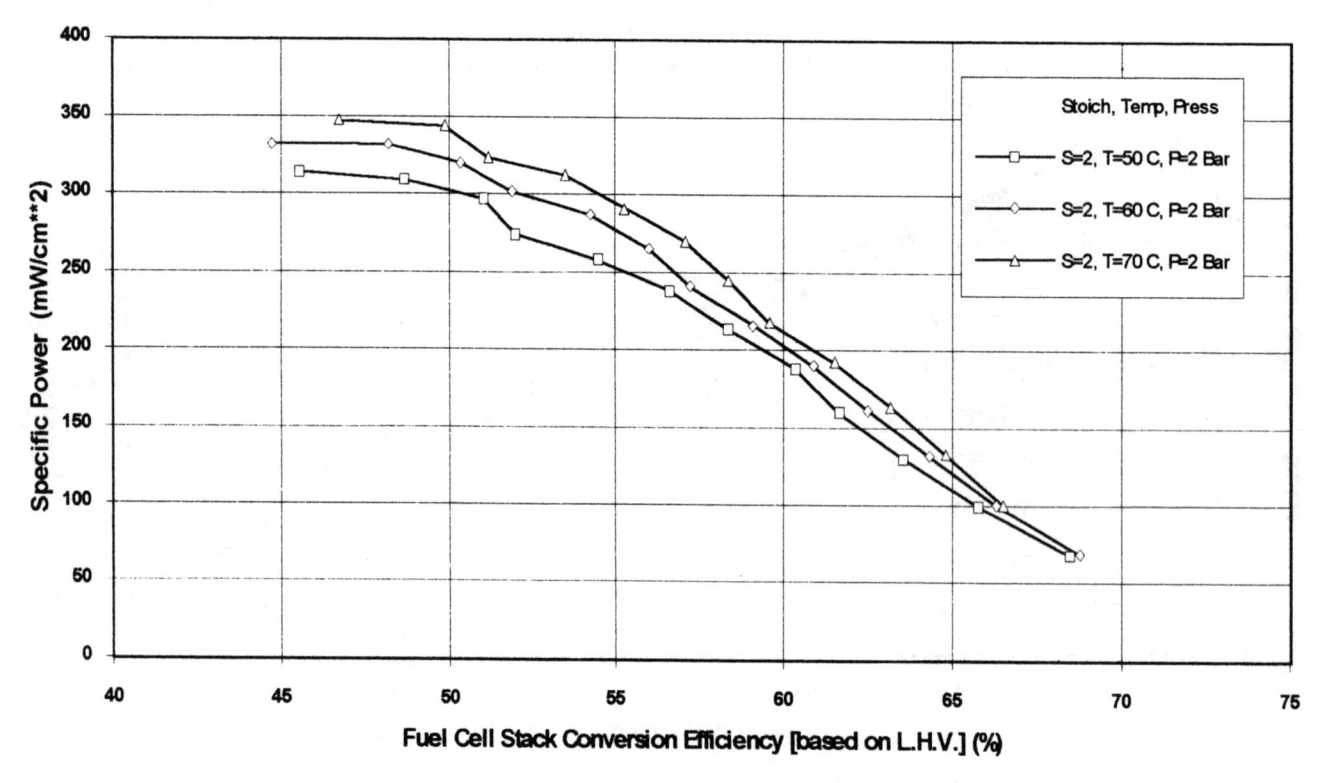

Figure 8. Specific Power as a Function of Temperature and Efficiency for Stoichiometry of 2.0, Pressure = 2 Bar

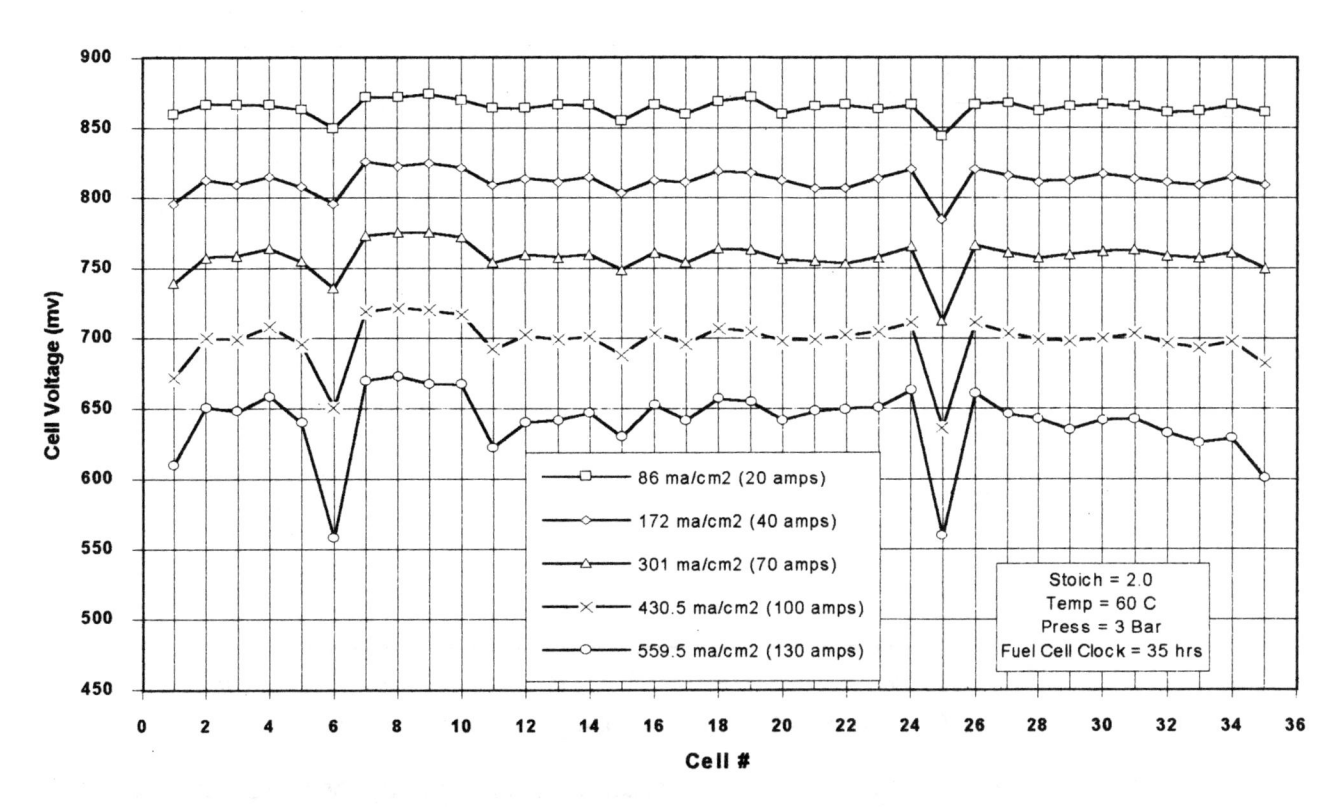

Figure 5. Individual Cell Voltages at Different Current Densities

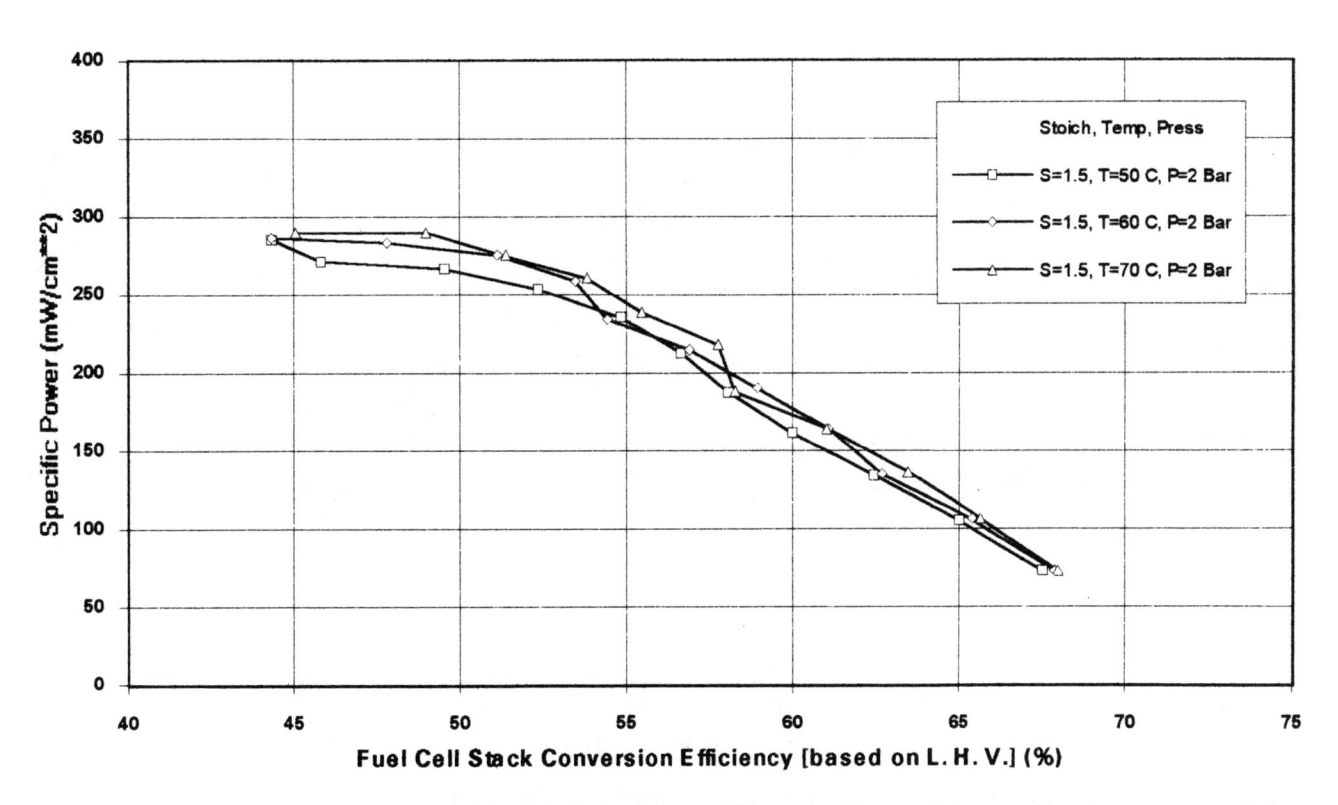

Figure 6. Specific Power as a Function of Temperature and Efficiency for Stoichiometry of 1.5, Pressure = 2 Bar

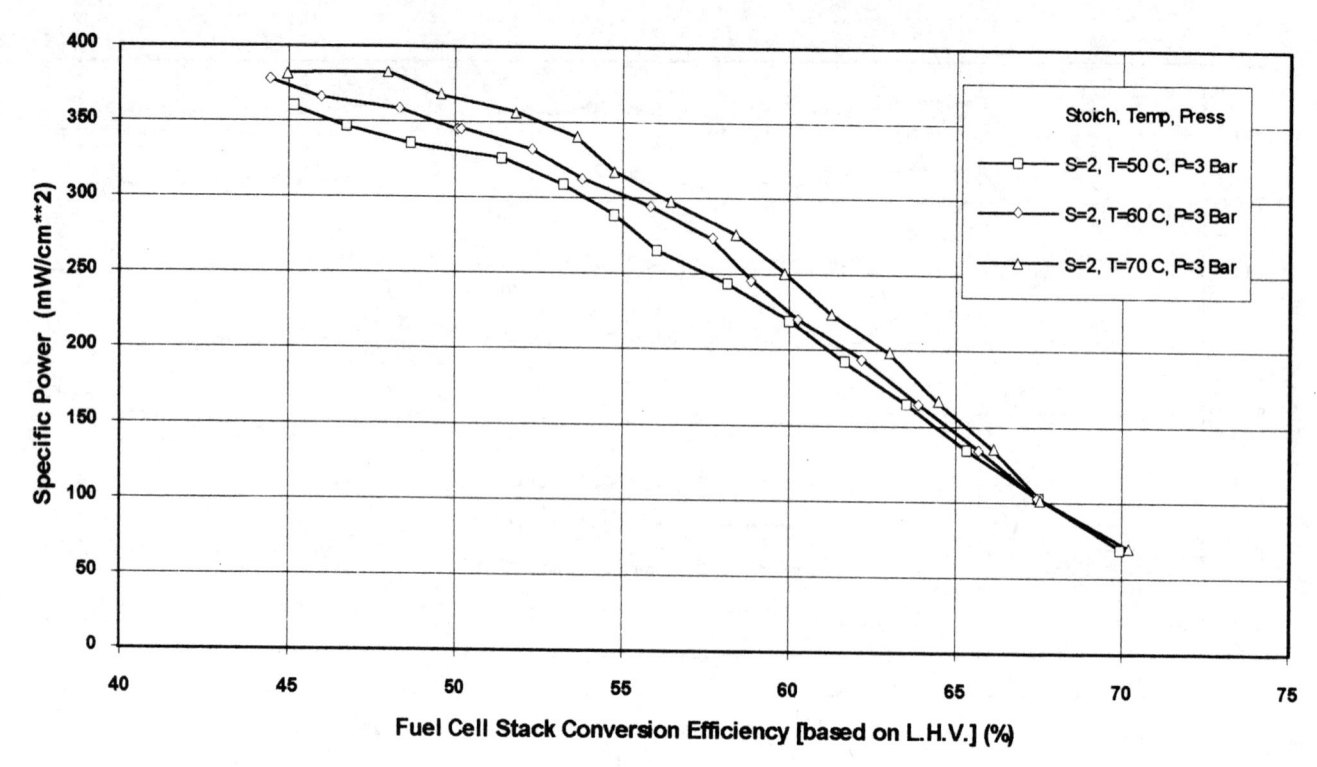

Figure 9. Specific Power as a Function of Temperature and Efficiency for Stoichiometry of 2.0, Pressure = 3 Bar

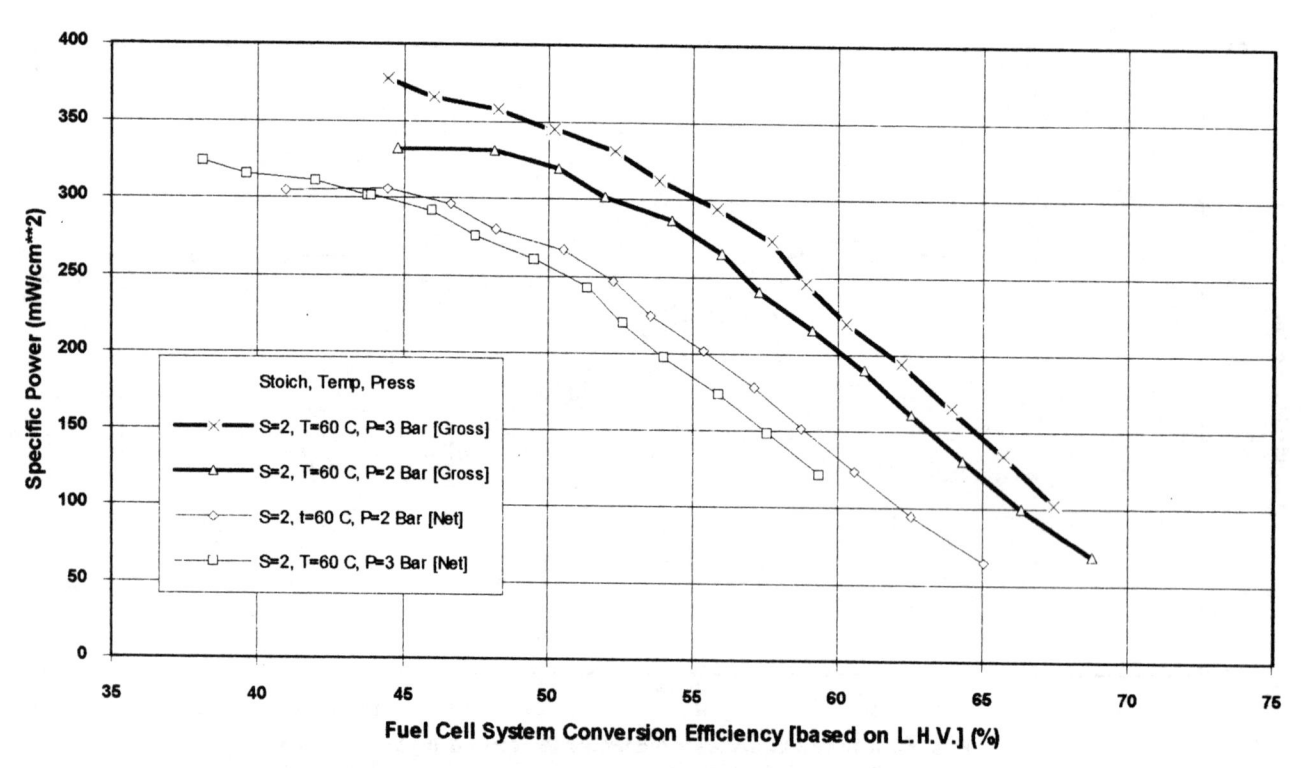

Figure 10. Comparison of Gross and Net Fuel Cell System Specific Power at Pressures of 2 and 3 Bar

940297

Power Quality Problems at Electric Vehicle's Charging Station

George G. Karady, Shahin H. Berisha, and Tracy Blake
Arizona State Univ.

Ray Hobbs
Arizona Public Service Co.

ABSTRACT

An area of electric vehicle (EV) racing which will require substantial study is the racetrack battery charging facility. During the 1993 third Annual APS Solar and Electric 500 race at Phoenix International Raceway (PIR) the generated current harmonics were monitored for various groups and types of battery chargers operating at the time. Measurements were taken on the low voltage side of the single phase supply transformer which supplied the charging station. Input current waveforms were recorded for 26 different chargers. The purpose of this paper is to show measurement results and statistically predict what the impact on the racetrack power system when large numbers of battery chargers are operated simultaneously. A statistical simulation was performed by adding randomly selected combinations of the digitized current waveforms which had been recorded at the racetrack charging facility. The THD was calculated for each iteration of the simulation. Measurements show that individual battery chargers generate distorted currents with THD in a range from 10% to 100%. Experimental measurement and simulations show that the total current generated by the simultaneous operation of several chargers is less distorted than individual charger currents. As the number of chargers increased there was a continual decrease in the THD of the combined currents.

1. INTRODUCTION

Electric vehicle technology is developing very fast. The electric car has a good chance of becoming a significant part of the transportation infrastructure in the near future. Although many technical details have yet to be resolved, automobile manufacturers all over the world are investing in research and development to be able to provide this alternative means of transportation.

Electric vehicles are widely promoted as pollution-free transportation which would reduce pollution levels in crowded cities. According to [1] the emission of carbon monoxide and other compounds would be slashed by more than 99% with an electric vehicle as compared to a gasoline powered vehicle in the same category.

There are several reasons why the EV has not become more prevalent as a means of transportation. A primary reason involves battery technology and battery charging systems. In general, charging a battery to optimum capacity is a slow process [1]. Charging some batteries produces gases which result in additional constraints on the charging environment. If the charging rate is too high overheating can be a problem for some batteries. Battery charging facilities will be constrained due to limitations of the electric power system. For example, to recharge a 20 kWh battery pack (a reasonable passenger car size) in 5 minutes from the 220 V line would require a current of more than 1000 A.

EV charging will have a great impact on electric utilities. Utilities will have to provide charging capacities as well as deal with the problem of power quality. Power quality issues are gaining in importance with the increased use of sensitive devices such as computers.

A literature review did not indicate previous work on harmonic generation at electric vehicle charging facilities. The 1993 APS Solar and Electric 500 electric vehicle race represented a unique opportunity to collect valuable data as there were a significant number of different types of battery chargers operating at the race. Due to non-sinusoidal currents generated by each charger, varying levels of harmonics will be injected into the power system. Input current waveforms were recorded for 26 different chargers at the race. The rms value and THD were calculated for each of the current waveforms. Due to the lack of sufficient data on cumulative currents theoretical simulations were performed to determine harmonics introduced into the system by the simultaneous operation of various numbers of chargers.

2. QUANTIFYING POWER QUALITY ASSOCIATED WITH BATTERY CHARGERS

Modern battery chargers are designed with power semiconductor devices (diodes, SCR's, and GTO's). Chargers used at the race varied from simple diode rectifiers to variable voltage, commercially available chargers which used switching circuits to control the charger waveforms. The process of rectification in the battery chargers will generate non-sinusoidal currents which, in turn, may generate a voltage distortion in the power system.

To evaluate the harmonic content of a distorted, non-sinusoidal current, a Fourier analysis of the current wave form was performed. From the results of the Fourier analysis the rms value of the current can be found from the equation:

$$I = \sqrt{\sum_i I_i^2} \quad (1)$$

where:
- I_i is the i th harmonic of the current

The total harmonic distortion (THD) can be used to represent the level of distortion due to the presence of harmonics:

$$THD = \frac{\sqrt{I^2 - I_1^2}}{I_1} \quad (2)$$

where:
- I is rms value of distorted current,
- I_1 is fundamental (60 Hz) harmonic of distorted current.

Power factor is the ratio of real to apparent power. With nonlinear loads, it is comprised of two components, displacement power factor and distortion power factor. Displacement power factor is the cosine of the displacement angle between the fundamental components of voltage and current. Distortion power factor is the result of harmonic current flows and can be calculated from the THD as follows:

$$pf = \frac{1}{\sqrt{1 + THD^2}} \quad (3)$$

Results for average, minimum, and maximum values are presented later.

3. DESCRIPTION OF A CHARGING STATION

The charging station at the APS Solar and Electric 500 was called the Electric Vehicle Refueling Station (EVRS) and was constructed specifically for the purpose of recharging electric vehicles. Power was delivered to circuit breaker panels inside the charging trailer from a single phase transformer located in the infield of Phoenix International Raceway. The panels provided 120 and 240 V to receptacles mounted on the end of the trailer. The circuit breakers in the panels provided the necessary fault protection. Three types of receptacles were installed: 110V at 20A, 220V at 20A, and 220V at 50A.

The minimum and maximum ratings of the 56 battery chargers inspected before the race was 110 V/20 A ac and 220 V/50 A ac respectively. Sufficient numbers of receptacles were provided to eliminate any waiting period.

The one line diagram of the charging station is shown in Fig. 1.

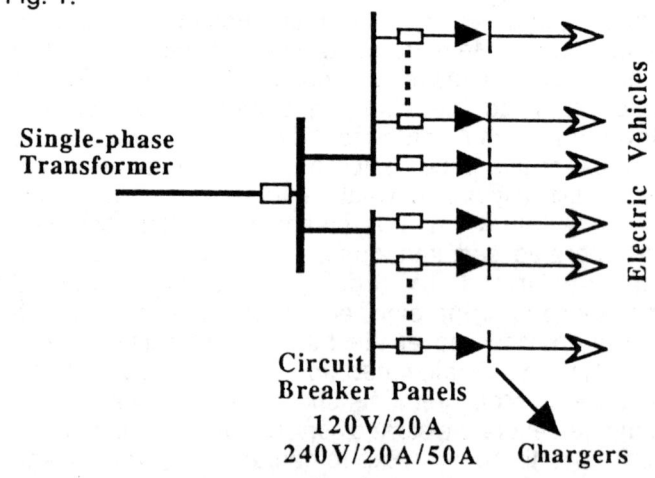

a. One line diagram

b. Relative locations of equipment
Figure 1. Charging station at PIR.

Vehicles entering the charging area were given a location and receptacle for charging on a first-come, first-served basis. Cars were escorted to the charging location and a ground continuity test was performed for safety reasons. Once the safety check was completed, a receptacle was selected (120V or 240V) and the charger was plugged in. On some vehicles the chargers were located inside the vehicle, while on others it was located outside. Some vehicles had two chargers, one for each of two sets of batteries. As many as 20 different chargers were operating simultaneously.

4. MEASUREMENT PROCEDURE AND ANALYSIS HARDWARE

The goal was to record a variety of different charger input ac waveforms and a sufficient number of current waveforms to be able to perform the statistical analysis. Measurements were taken while the vehicle was in the charging mode. The ac current to the charger was initially viewed with a digital oscilloscope to determine optimum recorder speed. Analog data were then stored on the FM analog data recorder. Configuration of the recording hardware is shown in Fig. 2a. If an observed waveform was similar to a previously recorded current it was not recorded. Current waveforms for 26 of the chargers were recorded from a total of 56 cars at the race, which allowed for a good statistical analysis. The data acquisition and analysis system is shown in Fig. 2b.

a. Recording hardware

b. Data analysis system

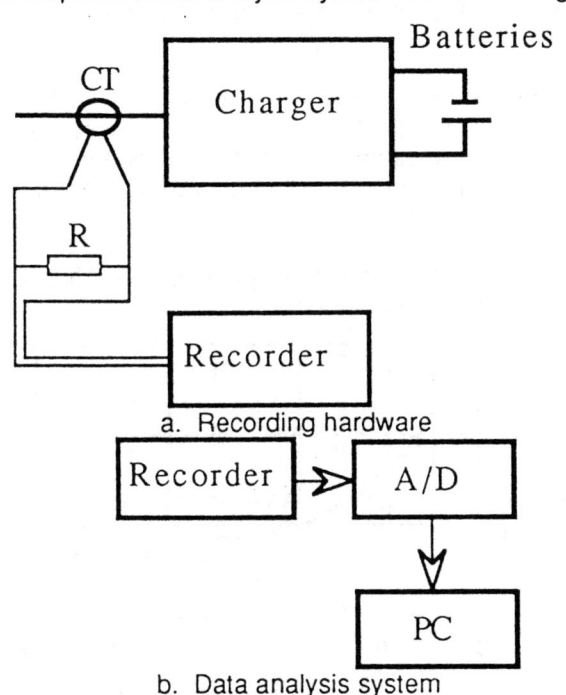

Figure 2. Recording and analysis system

The overall system has three main components: a) sensor, b) recorder, and c) data processor. The sensor is a commercially available clamp-on current transformer with a frequency response up to 100 kHz. A resistor is connected at the secondary side of the CT to maintain an optimal input voltage range to the recorder. The recorder used was a FM multi-channel analog tape recorder with a frequency response of DC to 20 kHz. Back in the laboratory the recorded data were digitized (A/D converter) and analyzed with an IBM computer.

5. MEASUREMENT RESULTS

Sixteen commercially available and 4 proprietary chargers were used during the race. Representative input current waveforms for three different chargers are shown in Fig. 3a, b, and c.

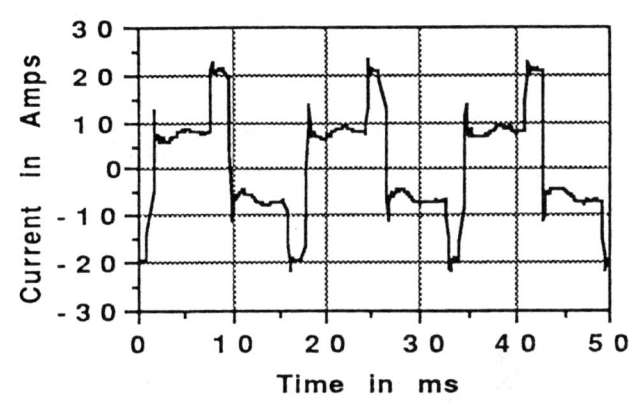

a. Battery charger on vehicle number 13

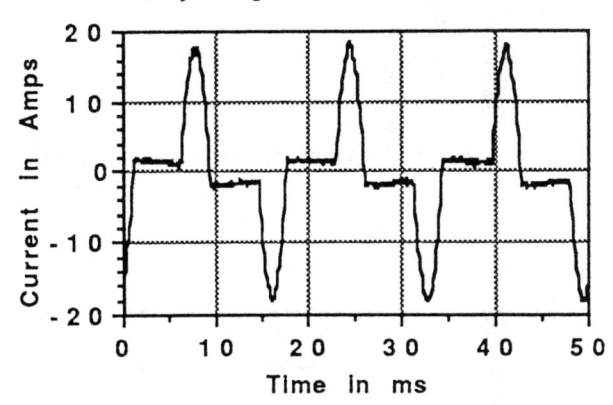

b. Battery charger on vehicle number 25

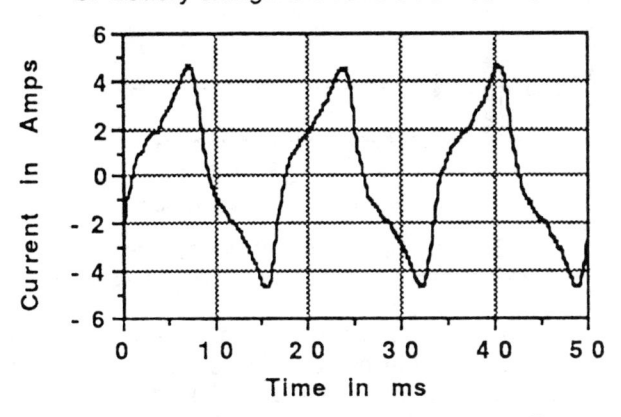

c. Battery charger on vehicle number 65

Figure 3. Representative current waveforms

Fig. 3a and 3b are examples of input current waveforms which contain high levels of harmonics. Fig. 3c is an example of a charger current with a much lower level of generated harmonics. The variety of the current waveforms shows that many different rectification techniques were employed in the battery chargers. These various techniques will produce varying levels of current harmonics as seen in Fig. 4.a b and c.

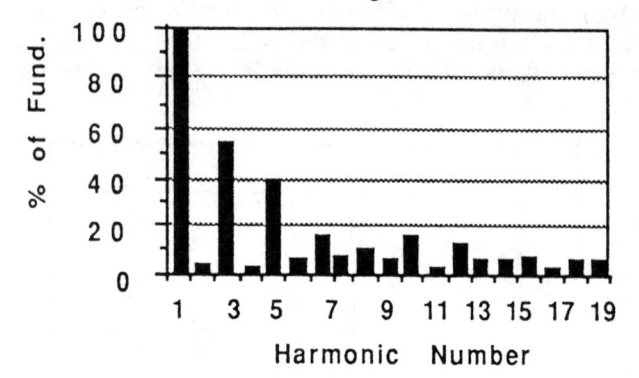

a. Battery charger on vehicle number 13

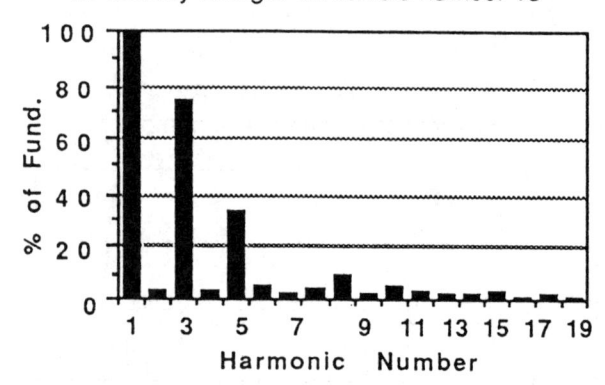

b. Battery charger on vehicle number 25

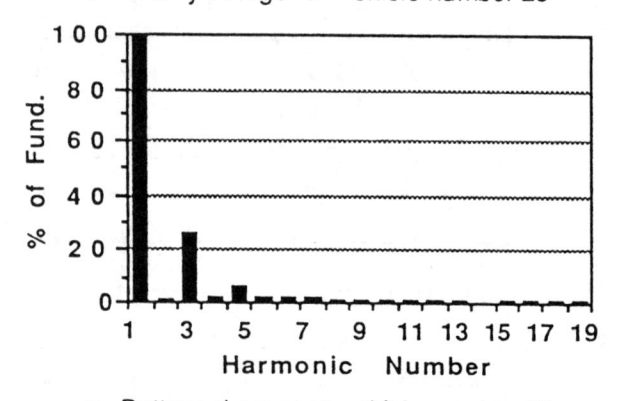

c. Battery charger on vehicle number 65
Figure 4. Representative frequency spectrums

The plots seen in Fig. 4 show that all chargers generate significant current harmonics. Current waveforms were analyzed up to the 50th harmonic (3 kHz) but the plots show frequency components only up to the 20th harmonic (1.2 kHz). Frequencies above the 20th harmonic were found not to be significant in the THD calculation. All chargers have at least one harmonic which is greater than 20% of the fundamental; Fig. 4b shows a charger with a third harmonic whose amplitude is greater than 75%. Many chargers also had higher harmonics above the fourth, at significant amplitudes. Fig. 4a shows a charger which has 5th, 7th, 9th, 11th, and 13th harmonics at an amplitude greater than 10% of the fundamental. Fig. 4c illustrates the reduced level of harmonics generated by the charger whose waveform was seen in Fig. 3c.

From the harmonic data for all of the battery chargers the minimum and maximum rms values of the first 50 harmonics were obtained. Fig. 5 is a plot of these minimum and maximum values.

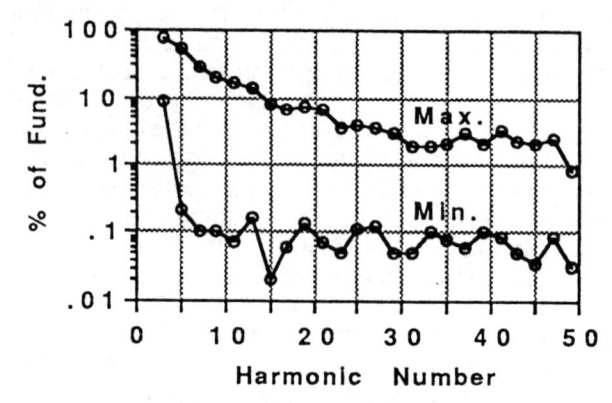

Figure 5. Upper and lower limits of odd harmonic amplitude

The minimum value for the third harmonic is 10% and the maximum is 78.5%. The maximum observed for harmonics in the range of 5 to 47 is between 2% and 52%. The minimum level observed for harmonics between 5 and 49 is below 0.2%.

Fig. 6 is a plot of the average values of harmonic amplitude for the 26 tested chargers.

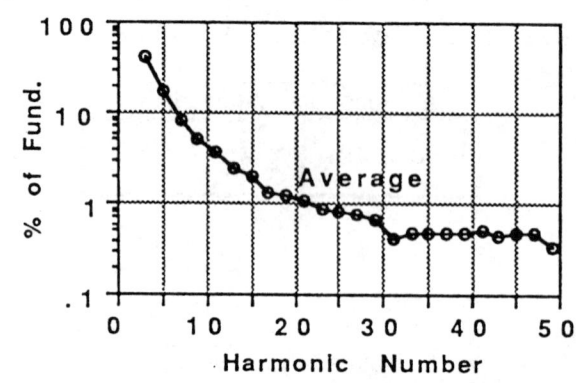

Figure 6. Average rms values of the odd harmonics.

The plot in Fig. 6 shows the average rms value of the odd harmonics less than 21 is greater than 1%. The harmonics greater than or equal to 21 have an average amplitude which is less than 1%.

Fig. 7 shows cumulative distribution of the THD for the chargers. Approximately 50% of the chargers have a THD of 50% or greater and only about 20% of the chargers have a THD of 20% or below. The average THD of all tested battery chargers is 50.1%

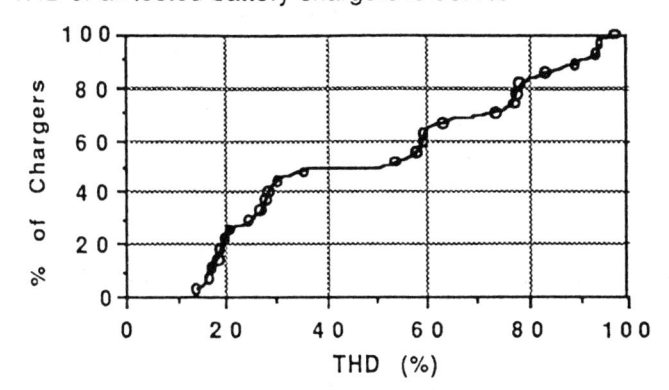

Figure 7. THD of the currents for each battery charger

Fig. 8 is the cumulative distribution of the distortion power factor for the chargers which was calculated using equation 3.

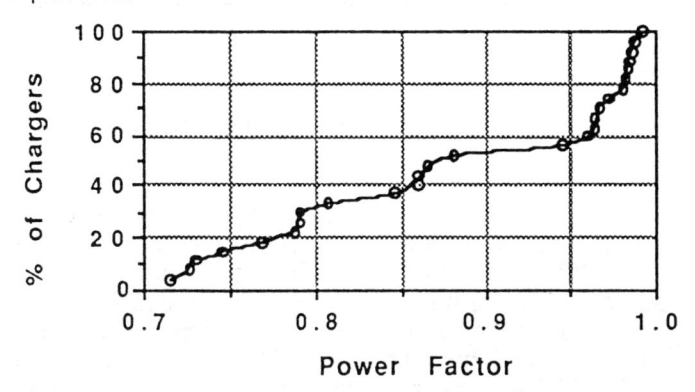

Figure 8. Cumulative distribution of the calculated distortion power factor

The minimum and maximum distortion power factor are 0.72 and 0.99 respectively. 40% of the chargers tested have a distortion power factor which is less than 0.85 and the average value of distortion power factor is 0.88.

6. EFFECT OF SIMULTANEOUS OPERATIONS OF THE CHARGERS

The waveforms and frequency spectrum plot shown in Figs. 3 and 4 are representative of individual chargers. The cumulative effects of multiple chargers operating simultaneously will now be discussed. Total charger current was measured at the two circuit breaker panels in the charging station at the APS Solar and Electric 500 race. The cumulative current data were taken on the last day of the race at 30 minute intervals. Fig. 9 shows typical cumulative current waveforms at the breaker panels.

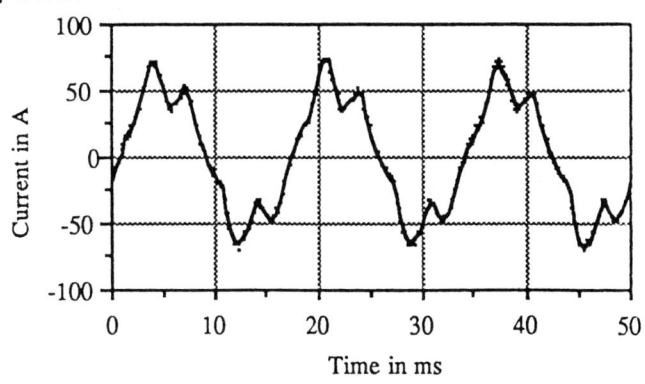

a. Cumulative current measured on circuit breaker I

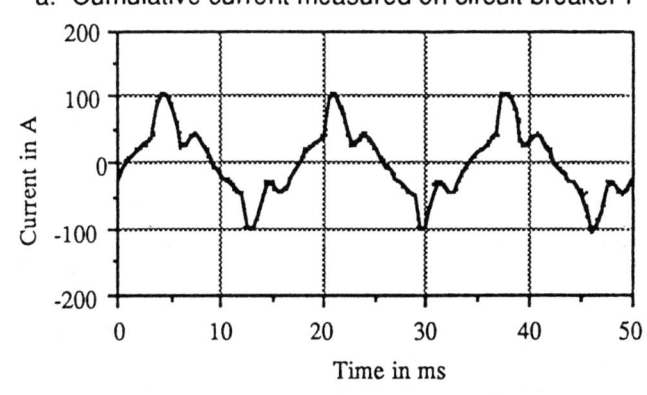

b. Cumulative current measured on circuit breaker II
Figure 9. Cumulative current waveforms

The currents shown in Fig. 9 are less distorted than many of the individual charger currents. The differences in the two waveforms are due to the varying numbers and types of chargers simultaneously connected to the two panels at the time of the measurements.

Calculated values of the THD for cumulative currents, which were measured at 5 different times during the day, are given in Fig. 10.

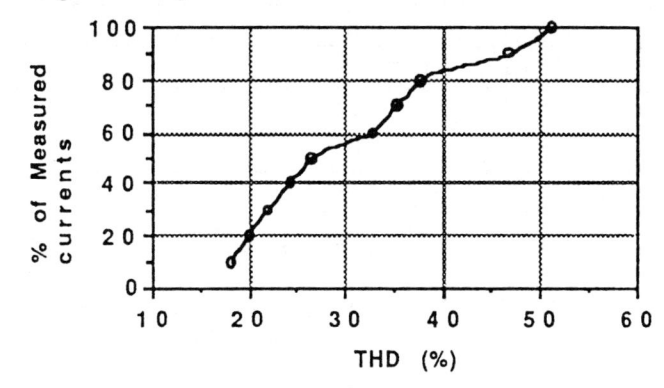

Figure 10. THD of cumulative currents

The time variation of the THD of cumulative currents is given in Table 1.

Table 1. THD of current at two circuit breakers

Time	THD Breaker 1	THD Breaker 2
9:00 am	18.1	35.2
9:30 am	24.2	46.7
10:00am	21.8	37.6
10:30am	19.9	51.1
11:00am	32.8	26.5

Depending on the number of chargers operating at the time, the THD of the cumulative currents varied from 18.1 (panel I, 9:00 a.m.) to 51.1% (panel II, 10:30 a.m.). The average values of the THD in Table 1 is 31.4%. This value is 37% lower than the 50.1%, the average THD of all chargers, and an indication that individual charger currents add in such a way as to reduce the level of distortion of the cumulative current.

Due to the small amount of experimental data on cumulative chargers a Monte Carlo simulation was performed to study the THD to be expected when different chargers are operating simultaneously. This type of statistical simulation can also provide insight into possible future power quality problems associated with charging stations. Initially, all 26 currents were scaled to produce waveforms with approximately the same rms value. The simulation was performed by varying the number of randomly chosen chargers operating at the same time. The simulation began with 5 chargers; this number was increased to 25 chargers in steps of 5. 500 simulations were performed for each case. The total current and corresponding THD were calculated. A typical histogram showing distribution of the THD for 500 simulations is shown in Fig. 11.

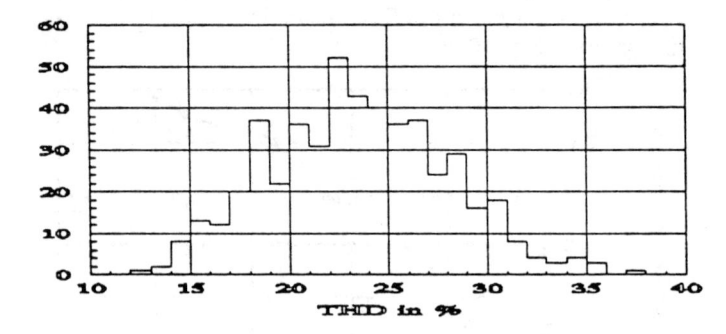

Figure 11. Typical THD histogram developed based on 500 simulations for 20 chargers

The mean of the quasi Gaussian distribution is 32.0%. Table 2 documents the statistical data for the simulations.

Table 2. Statistical parameters obtained by simulation

Nr. of Charg.	Min. THD	Med. THD	Avg. THD	Max. THD	St. D. THD
5	3.4	25.4	28.7	112.4	16.0
10	5.7	23.5	24.7	64.9	9.8
15	6.2	23.4	23.5	44.4	6.7
20	12.4	22.7	23.0	36.9	4.5
25	20.2	23.1	22.8	25.2	1.6

Table 2 shows that with an increasing number of chargers operating at the same time the THD decreases. Graphical representation of this variation is shown in Fig. 12.

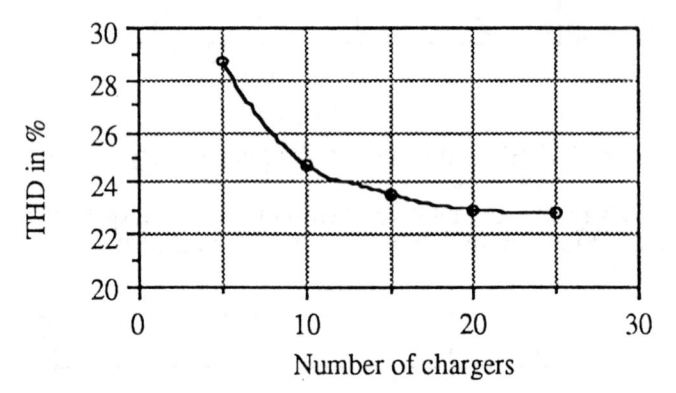

Figure 12. Variation of the THD from Monte Carlo simulation.

Table 2 and Fig. 12 clearly show that as the number of chargers increases the THD decreases to a minimum of about 22%. Limits on the THD presented in IEEE Standard 519 are based upon the ratio of available short circuit current to load current at the point of common coupling (circuit breaker panel). The standard limits the THD to 5.0% if the available short circuit current/load current ratio is less than 20. The ratio at the charging station was below this number, but Fig. 12 shows the THD from the simulations to be much greater than 5%.

CONCLUSION

Based on the results of AC current measurements of battery chargers at the charging station at the APS Solar and Electric 500 race, the following conclusions can be drawn.

1. The AC current waveforms demonstrate the diversity in charger operation and technologies present at the APS Solar and Electric 500. All of these chargers generated non-sinusoidal periodic waveforms with varying levels of distortion.

2. Battery chargers used at future racing events must have tighter requirements regarding the distortion they introduce into the power system. This is evident from results of individual charger evaluations at the 1993 APS Solar and Electric 500 presented in this paper. After testing, 26 battery chargers were found to have a minimum and maximum THD value of 13.6% and 97.6% respectively. The average THD generated by the chargers was 50.1%.

The minimum and maximum values of distortion power factor were 0.72 and 0.99 respectively. The average distortion power factor produced by the battery chargers was 0.88. Fifty percent of the battery chargers had a power factor less than 0.9, and twenty five percent of the battery chargers produce a distortion power factor lower than 0.75.

3. Odd harmonics were the predominant frequencies present in the input current waveforms, but some even harmonics were present at significantly lower levels. Some chargers had frequency components above 1.2 kHz (20th harmonic) but there were insignificant in the power quality analysis.

4. Increasing the number of chargers in operation reduces the THD of the total current. This result was verified statistically by using a Monte Carlo simulation.

5. To minimize power quality problems associated with charger operation:

   a) a specification must be drafted which will limit distortion of the input ac current generated by individual chargers. Application of harmonic reduction filters would also improve power quality.
   b) under racing conditions a specification must be drafted to require that a minimum number of chargers be operational at any given time.

REFERENCES

[1] IEEE Spectrum, Electric Vehicles Nov. 1992.
[2] EPRI Report TR-101322: Status and Trend Assessment of Advanced Battery Charging Technologies, prepared by San Jose University Nov. 1992
[3] G.T. Heydt: "Electric Power Quality", 1991
[4] W. Shephard, P. Zand: "Energy Flow and Power Factor in Non-sinusoidal Circuits", Cambridge University Press, London 1979
[5] IEEE, "Guide for Harmonic Control and Reactive Compensation of Static Power Converters," ANSI/IEEE Standard 519, 1981.

940298

Low Frequency Magnetic Field Generated at Electric Vehicle's Charging Station

George G. Karady, Shahin H. Berisha, and Mukund Muralidhar
Arizona State Univ.

Ray S. Hobbs
Arizona Public Service Co.

ABSTRACT

The electric vehicle is one alternative form of transportation that has gained importance over the last few years due, in part, to pollution-free operation. Electric vehicles in their charging mode generate magnetic fields. Over the last few years these fields have been investigated for possible health hazards. A study was undertaken to measure and evaluate the magnetic field generated within a charging station. The 1993 Arizona Public Service Solar and Electric 500, a race for solar and electric vehicles held at Phoenix International Raceway, offered an excellent opportunity to collect data on the magnetic field generated within a charging station. Quantification of the magnetic field is necessary to predict the level of the field, the location where the maximum field is generated, distribution of the field within the charging area, etc. An electric vehicle requires both periodic and fast charging times. These factors necessitate building a large number of charging stations in the future, each with sufficient capacity to service a large number of vehicles simultaneously. Therefore, the requirements of a charging station were also studied. The magnetic field was measured inside and around the charging trailer, along the charging cords and at four points around each of the electric vehicles in their charging mode. This paper characterizes the overall magnetic field generated in the charging station. The maximum field magnitude measured is 64.37 mG at a distance of 50 cm from the circuit breaker panels. Results show that a low frequency magnetic field is generated. The frequency range for the field is 60 - 420 Hz.

INTRODUCTION

Electric vehicle (EV) technology has improved significantly in the last 20-25 years. Previous results [1] show that EVs in the charging mode generate significantly high magnetic fields both inside and outside the vehicle. EVs need periodic as well as fast rates of charging. An increasing number of EVs are being built and tested; production of a large number of EVs is expected in the near future. This will necessitate building a number of charging stations, each of sufficient capacity to service a large number of vehicles simultaneously. The magnetic field generated within a charging station must be measured., as the expected field inside a station is not known.

Since human exposure to magnetic fields may be significant, they are being investigated for possible health hazards. The charging station is an area where a lot of people are constantly exposed to the magnetic field for long periods of time. Hence it is very important to get a detailed knowledge of the field magnitude, distribution of the field, etc. It is also necessary to make measurements and know the field levels to find techniques to reduce the fields.

The field was measured at different locations within the charging station. Analysis of the data would not only help predict field levels in a future charging station but also help to design that charging station. Knowledge of distribution of the field level from measurements and any problems encountered while measurements are performed would aid the planning and layout of components comprising a charging station, such as arrangement and location of the receptacles, if power cords from circuit breaker panels to receptacles should lie underground, if cables require shielding, number of charging receptacles, type of power supply, switching circuit employed on the EV, how close vehicles can be parked side-by-side, etc. Electric utilities would also benefit from valuable data such as the rating of future charging stations, amount of harmonic distortion present in the currents, etc. With a magnetic field generated by a number of different of battery charger systems, the racing environment simulated a future charging station where 20 to 30 vehicles will be charged simultaneously. This paper characterizes the overall magnetic field generated within a charging station.

DESCRIPTION OF THE CHARGING STATION

The charging station was called the Electric Vehicle Refueling Station (EVRS) and was built specifically for the race. Figures 1a and 1b show the one line diagram and relative locations of equipment in the charging station respectively. Power was delivered to circuit breaker panels inside the charging trailer from a single phase transformer located in one corner of the charging area. Necessary fault protection was provided by these panels. Three types of single phase receptacles were provided: a) 110 V, 20 A, b) 220 V, 20 A, and c) 220 V, 50 A. A total of 54 receptacles were mounted on a platform in the front of the charging trailer as shown in Figure 1b.

The total number of vehicles in the race was 68. Average charging and waiting times depended on the rating of the charger and, therefore, the receptacle the vehicle was plugged into. Taking into account the different types of chargers, average charging time was from 6 - 10 hours and average waiting time was less than one hour. Because there were 54 receptacles and only 68 vehicles, there was not a problem of any vehicle waiting a long time to be charged and the charging station was able to meet the requirements of the all the EVs. Before the start of the race 26 battery chargers were inspected. The minimum rating was 110 V, 20 A and the maximum rating was 220 V, 50 A. A wide range of chargers were seen, ranging from sophisticated and commercially available to high school/home-made.

Vehicles were escorted to their charging locations and a safety check were performed. This continuity check, to ascertain that all metallic parts of the electric vehicle were tied to a common ground point, was repeated each time a vehicle was presented for recharging. Once the safety check was completed the battery charger was plugged into the proper receptacle and charged. Chargers were either built into the vehicle, or located outside the vehicle, on the ground. Some vehicles had two chargers for two sets of batteries. Up to 30 chargers were plugged in simultaneously. Because a variety of chargers were operating simultaneously, it was a complicated environment in which to characterize the magnetic field. Also, as seen from Figure 1b, cables from the receptacles to the vehicles were on the ground, bunched together between the two rows of vehicles. and close to the charger. This also made it difficult to separate the field, as the various sources of the field could not be isolated from each other.

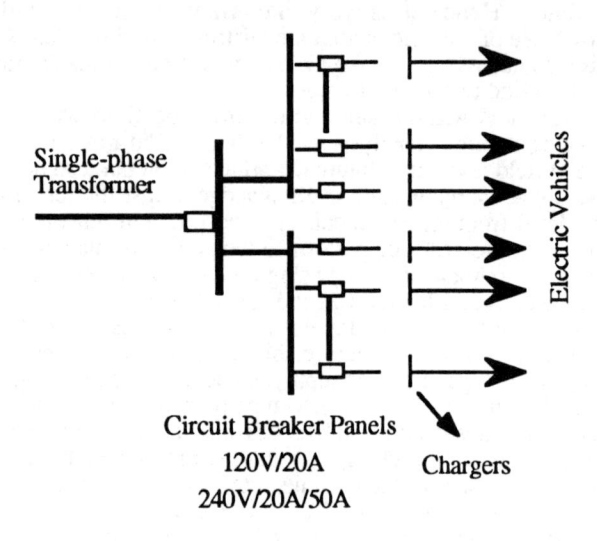

a. One Line Diagram.

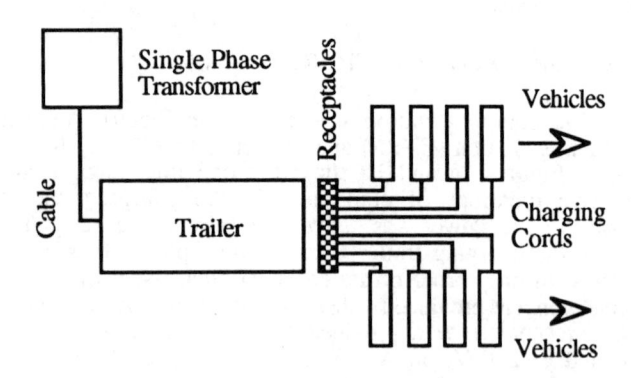

b. Relative Locations of Equipment.

Figure 1. Charging Station at PIR.

TYPES OF ELECTRICAL SYSTEMS

The EV has two electrical systems, the drive system and the charger system.

DRIVE SYSTEM - There are two types of drive systems: ac drives and dc drives. The ac drive systems use an inverter to convert the dc battery supply to ac supply to drive an ac motor. The dc drive systems use a chopper to change the level of the dc battery supply to drive a dc motor.

CHARGER SYSTEM - The charger system has a rectifier which converts the ac supply to dc supply to charge the batteries.

All measurements were made when the vehicles were in their charging mode so the drive systems did not affect the magnetic field. Switching elements in the battery charger generate nonsinusoidal periodic currents. Thus, EVs generate magnetic fields which have frequency components other than 60 Hz. The frequency components of the field depend on the switching circuit of the charger; the higher the distortion in the current is, the higher the maximum field will be. The total current is the sum of all the individual currents to the chargers. Thus, depending on the number of EVs and the switching circuit of the charger, the total current could have various levels of distortion. This quantity, the total harmonic distortion (THD), is very important in determining power quality and has been dealt with in [2]. From the current waveform and the switching circuit present, it is possible to decide which charger would be the most appropriate to use over a long period of time. Future requirements for the charging station can be estimated from the number of vehicles present and the approximate amount of charging time required.

MEASUREMENT METHOD

Characterization of the magnetic field generated by the charging station requires measurement of the field at various locations within the charging area. The functional block diagram for measurement and acquisition of the magnetic field is shown in Figure 2.

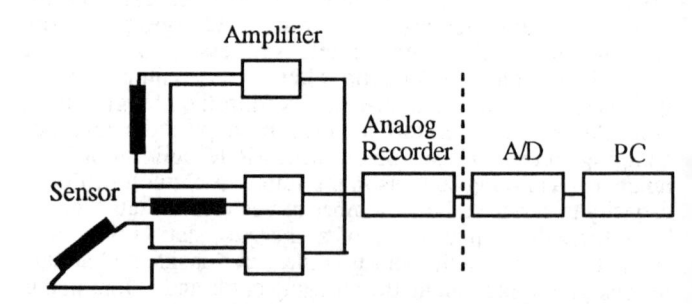

Figure 2. Magnetic Field Measurement System.

The various instruments used in the measurements are described below:

SENSOR - Three magnetic field sensors are orthogonally mounted on a 15x15x15 cm wooden cube. The sensors are of the magnetometer type; their output voltage is proportional to the X, Y and Z components of the magnetic field. Each sensor output is fed to a preamplifier. The output of the preamplifier is a maximum of 2 V for the three measurement ranges of the sensor: 20, 200, and 2000 mG. The frequency response of the three sensors is dc to 440 Hz.

ANALOG RECORDER - The analog recorder has seven input channels and measures signals in the frequency range from dc to 20 kHz. The upper frequency limit varies from 400 Hz to 20 kHz depending on the speed of the tape. The recorder will store a huge amount of data on a magnetic tape.

A/D CONVERTER - A digital storage oscilloscope with a 10 bit A/D converter digitizes the analog signal. Although the signal is sampled at a rate of 1.2 kHz, only the first 10 harmonics of the frequency spectrum are considered.

PC - Digitized data are transferred to a PC using the data acquisition software GURU. The parameters characterizing the magnetic field are then evaluated using the PC version of MATHCAD, a general purpose calculation program.

EVALUATION METHOD

The field can be measured either in time domain or frequency domain at any given point. In the present work, measurements were performed in the time domain. At each point, the field can be represented either by a time variable vector or an amplitude time function. The parameters evaluated were: the peak and rms values, total harmonic distortion, and significant frequency range, etc. To assure easy comparison of the field at different locations the rms value including the dc component, the rms value excluding the dc component and the total harmonic distortion (THD) for both cases were calculated. The rms value defines the strength and the THD defines the frequency content of the field.

The magnetic field data were recorded in a multi-channel analog tape recorder. A digital oscilloscope was used to monitor output signals from the sensor. After data collection the analog signals were digitized using the digital oscilloscope. Digitized data were transferred to the computer for processing using the GURU data acquisition software.

For evaluation of the data, a calculation method was developed and implemented using MATHCAD, a general purpose calculation program. The evaluation method is presented in [3].

Each field component is represented by a set of digitized data containing N values. The time for each value of the voltage is the product of the sample number (i) and the sample interval Δt. The magnetic field is calculated by numerical integration. The calculation is done for all three coil voltages and produces the three magnetic field components. The frequency spectrum of the derivative of the magnetic field is calculated by FFT analysis. The magnetic field is calculated by dividing each component of the derivative by the corresponding angular frequency. The rms value and the THD are calculated from the frequency domain data. The evaluation procedure is explained in detail in [4].

MEASUREMENT LOCATIONS

Figure 3 shows measurement locations inside and around the charging trailer. Point CT1 is located outside the trailer, 50 cm above ground level and 50 cm from the supply cable to the charging trailer -- the cable carrying the cumulative current supplied to the charging station. Points CT2 and CT3 are located inside the trailer, 50 cm above the trailer floor and 50 cm from the two circuit breaker panels. Each circuit breaker panel supplies current to half the receptacles.; the currents in the two circuits should be balanced. Points CT4 and CT5 are located outside the trailer, 50 cm above ground level and 50 cm from the panel containing the receptacles.

The magnetic field at point CT1 is generated primarily by the total input current to the charging station. Fields at points CT2 and CT3 are generated by the two currents through the circuit breaker panels,. Fields at points CT4 and CT5 are generated by the ac currents of the battery chargers connected to the receptacles.

Figure 4 shows measurement points along the charging cables, which were bunched together. Each point was 50 cm above ground level. The cables were bunched most closely together near the receptacles and decreased from the receptacles to point A as shown in the figure. The magnetic field measured along the charging cables is generated primarily by the ac currents of the battery chargers. The strength of the magnetic field at these points depends on the number and type of the chargers connected at the time of measurement. Vehicles were parked in two rows along the charging cables. People involved in the race were always present in the area where measurements were performed and, therefore, constantly exposed to the magnetic field.

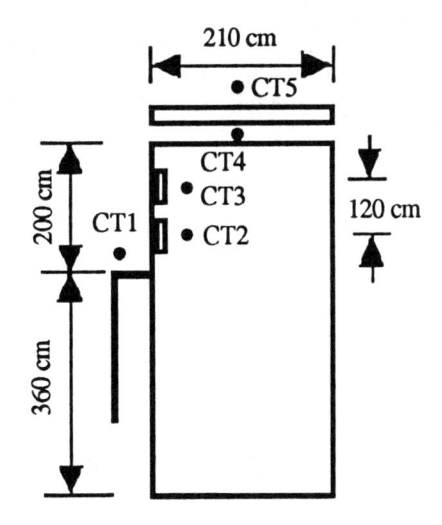

Figure 3. Measurement Points Inside and Around the Charging Trailer.

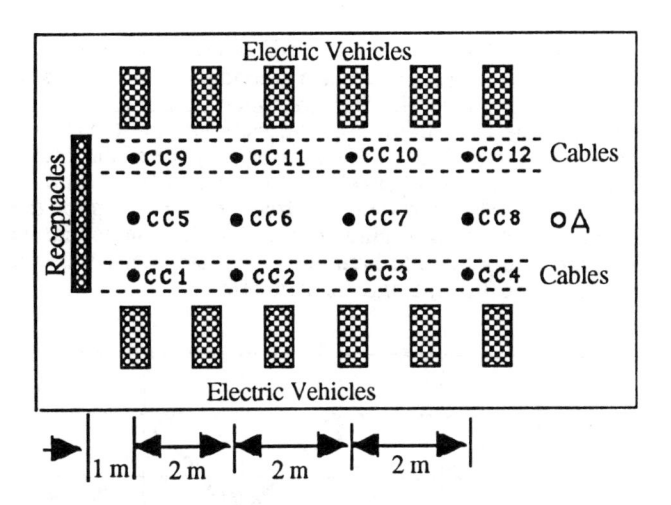

Figure 4. Measurement Points Along the Charging Cables.

Figure 5 shows measurement points around the EV. The field was measured at four points around each vehicle. All points were located 50 cm above ground level and 50 cm from the vehicle. Because the gap between adjacent vehicles was small, it was impossible to avoid the influence of the field generated by adjacent vehicles while measurements were obtained for a particular vehicle. Consequently, the magnetic field at the points around one vehicle is not necessarily generated by the electric system of that vehicle alone.

RESULTS OF MAGNETIC FIELD MEASUREMENTS

Throughout the four day period, a large number of frequency magnetic field measurements were taken around each of electric vehicle in its charging mode. Spot measurements at two or three locations were initially performed using a sensor with a frequency range from 30 Hz to 300 kHz. No dominant frequency components higher than 420 Hz were found. Evaluation of ac currents generated by 27 tested chargers [2] shows that each generates distorted periodic waveform currents. The 60 Hz frequency component was dominant in all the currents. The average value of the THD of these currents is 50.1 %. The frequency spectrum of these currents contains only low frequency components, hence, the EV and battery charger generated magnetic field also contains only low frequency components. The other major source of the magnetic field generated is the dc current from the chargers to the batteries. Because the dc current is much higher in magnitude than the ac current, the major portion of the magnetic field in the charging station will be due to the dc current. Large in magnitude, the dc field is still only about 10 % of the earth's field.

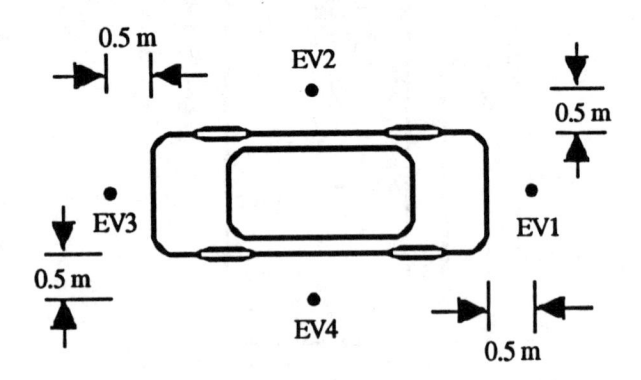

Figure 5. Measurement Points Around the Vehicles.

INSIDE AND AROUND THE TRAILER - Figure 3 shows various locations inside and around the trailer. Because the number of plugged-in chargers varies from hour to hour, currents in the two circuit breakers vary. Figure 6 shows a typical waveform of the current measured in one of the circuit breakers; the current is periodic and has a significant amount of distortion. The corresponding frequency spectrum of the current is shown in Figure 7; the 60 Hz component is dominant. Only the odd harmonics are dominantly present, particularly the 3rd, 5th, and 7th harmonics.

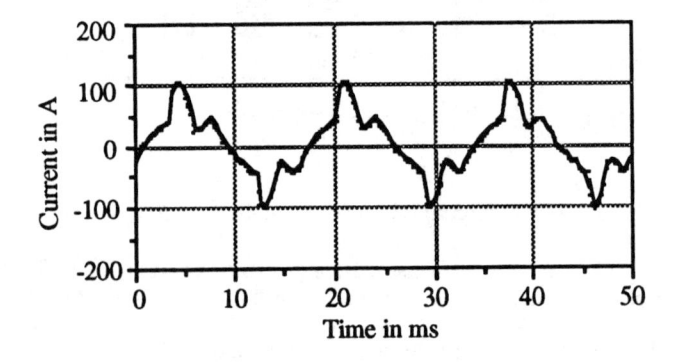

Figure 6. Current Waveform in Circuit Breaker.

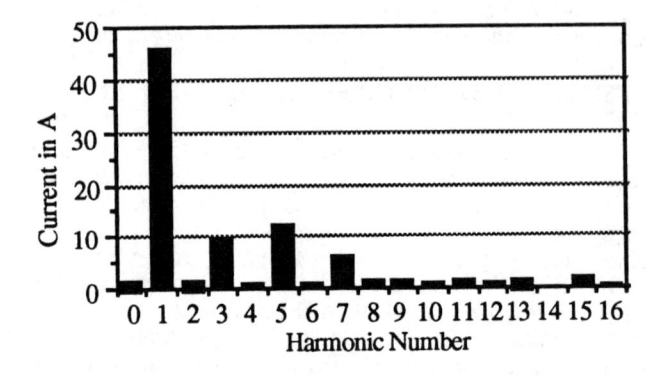

Figure 7. Frequency Spectrum of the Current.

Table 1 shows results of the magnetic field at the five points inside and around the trailer. Each measurement point is located in the region where there is no significant dc magnetic field present. The THD, with and without the dc field, is almost the same for locations CT1, CT2, and CT3. Point CT4 shows some deviation with and without the dc component. As expected, the highest magnetic field is measured at point CT1, located outside the trailer and generated by the supply cable to the two circuit breakers. Because the circuit breaker panels are not equally loaded there is a difference in the fields measured at points CT2 and CT3. The most distorted field waveform is measured at point CT5, located in front of the receptacles. The dc component of the magnetic field is the main reason for the distortion from 60 Hz at point CT5. The dc component is generated by battery chargers which are charging the EVs located near it. If the dc component is eliminated, the most distorted magnetic field is measured at point CT2 and the least distorted magnetic field at point CT3.

Table 1. Magnetic Field Inside and Outside the Trailer.

Point	Rms mG	Dc mG	THD %	THD (dc=0) %
CT1	32.30	3.35	69.80	68.03
CT2	16.60	2.57	89.40	85.74
CT3	26.70	0.61	36.30	36.21
CT4	5.20	0.53	66.10	50.76
CT5	2.10	1.04	117.90	64.92

Typical magnetic field wave forms in the X, Y, and Z directions are shown in Figures 8 a, b, and c respectively. All three components of the magnetic field are periodic waveforms with significant distortion. The magnetic field components at the points close to the circuit breakers carrying the ac currents contains a very small presence of dc components. The maximum value of the resultant field of the three components is 64.37 mG.

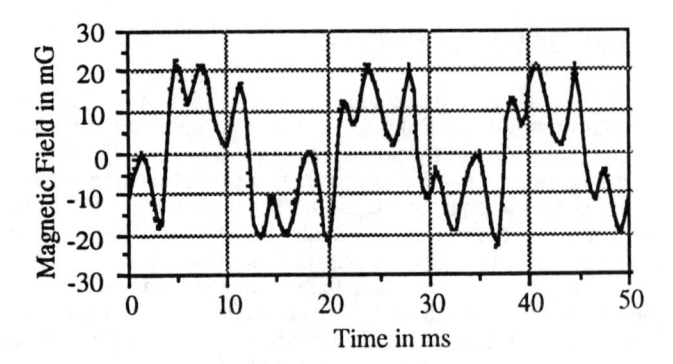

a. X - Direction.

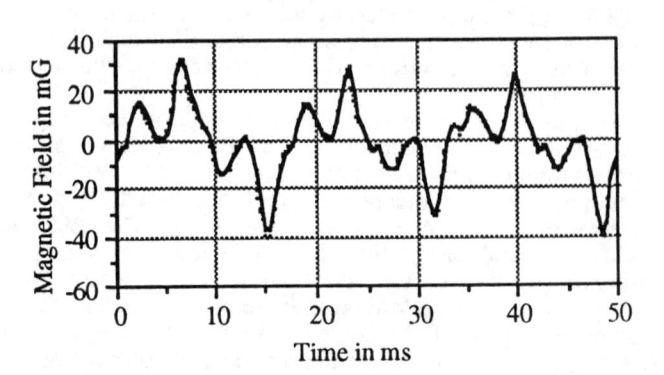

b. Y - Direction.

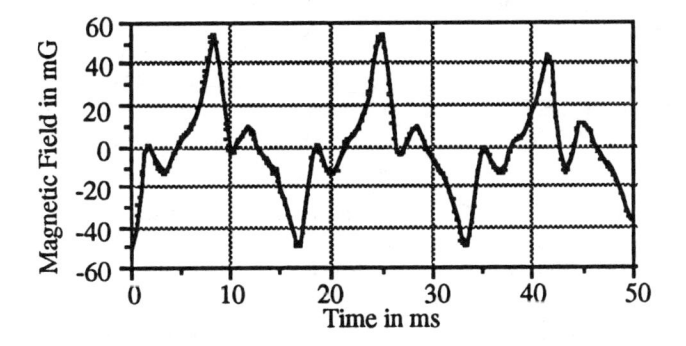

c. Z - Direction.

Figure 8. Components of the Magnetic Field Waveform at Point CT2.

Figure 9 shows the frequency spectrum of the magnetic field from Figure 8. As in the case of the current spectrum, odd harmonics are dominantly present. Significant presence of 3rd (33 % of fundamental) and 5th (29 % of fundamental) harmonics are found in the field spectrum. If the dc component is eliminated, the average value of the THD of the magnetic field in points inside and around the charging trailer is 61.12 %. This value of THD is close to the average value of THD (= 63.5 %) for the charging station as a whole.

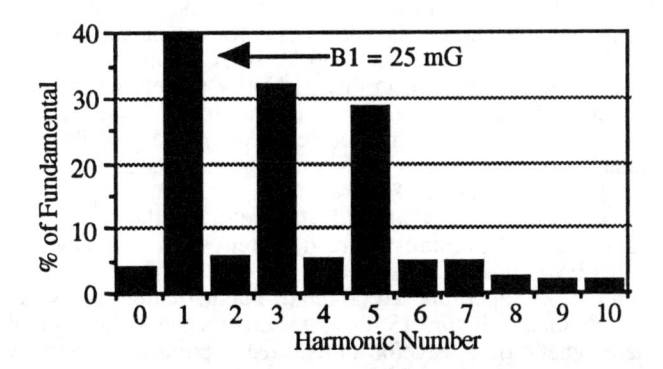

Figure 9. Typical Frequency Spectrum of Magnetic Field.

ALONG THE CHARGING CABLES - The magnetic field along the charging cables was measured at the points shown previously in Figure 4. A typical current waveform is presented in Figure 10.

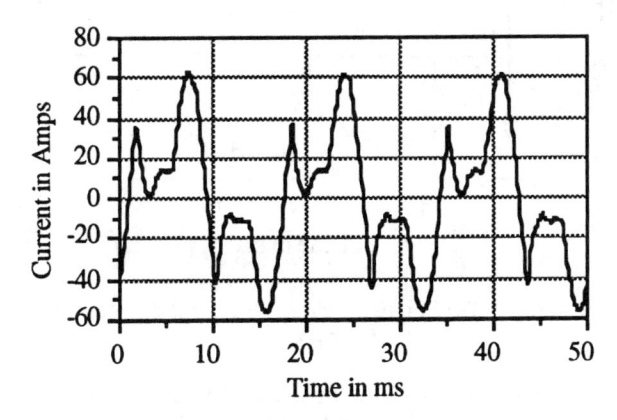

Figure 10. Current Waveforms Generated by Battery Chargers on Vehicle Number 33.

This magnetic field is generated by the ac and dc currents of the battery chargers. Since dc current to the charger is always higher in magnitude then ac current, in some cases a strong dc component was measured. Each charger generates a different current waveform depending on the switching circuit employed. The rating of vehicle number 33, shown in Figure 10, was 120 V, 27 A ac input and 120 V, 22 A dc output. The ac input current was measured in 26 battery chargers. Each generated distorted periodic ac currents. The average THD of the ac current to the chargers is 50.1 % [2].

Figures 11a and b show the cumulative distribution of the magnitude and the THD of the magnetic field respectively, along the charging cables. In Figure 11a, 80 % of the measurement points have magnitudes less than 20 mG. These points are furthest away from the trailer; the cables are not as closely bunched as they are near the trailer. The measured field is very distorted as shown by the high values of the THD in Figure 11b. The THD for this case has values similar to that of the charging trailer. If the dc magnetic field is eliminated the average value of THD is 51.5 %, very close to the calculated THD of the input ac currents to the battery chargers. Figure 11b shows that 40 % of the measurement points have a THD greater than 50 %.

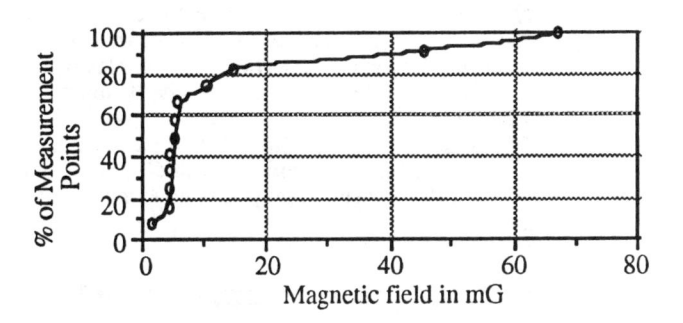

a. Magnetic Field

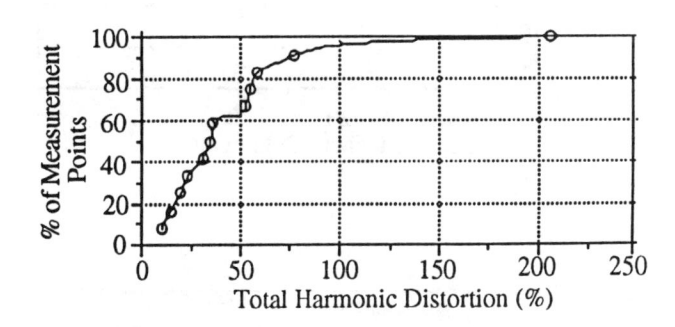

b. Total Harmonic Distortion

Figure 11. Cumulative Distributions of the Magnitude and THD of the Magnetic Field Measured Along the Charging Cables.

Waveforms similar to those plotted for the trailer show that the X, Y, and Z components of the magnetic field along the charging cables are distorted and also contain a significant amount of dc component. The frequency spectrum of the total magnetic field of the three components is shown in Figure 12. This figure shows the strong presence of the dc component which has the same value as the fundamental. The 3rd, 5th, and 7th harmonics are also significant.

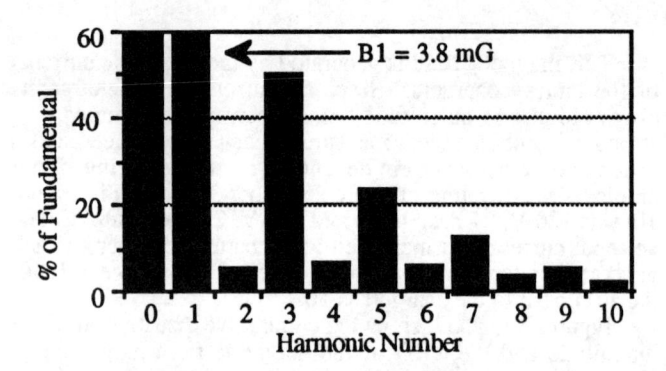

Figure 12. Frequency Spectrum of Magnetic Field Along the Charging Cables.

AROUND THE VEHICLES - The magnetic field was measured at four points around each of the 30 electric vehicles in their charging mode. Figure 5 shows the measurement locations. This magnetic field is generated by the input ac current and output dc current of the chargers. In certain vehicles the charger was located inside the vehicle (on-board).

When the battery charger is on-board, the magnetic field outside the vehicle has a smaller dc component because cables carrying the dc current are routed within the vehicle itself. When the charger is off-board, that portion of the cables which carry the dc current is outside the vehicle and the dc magnetic field is more dominant.

Figures 13a and b show the cumulative distributions of the magnetic field measured around the vehicles with and without dc. 90 % of the measurement points have magnitudes less than 20 mG.

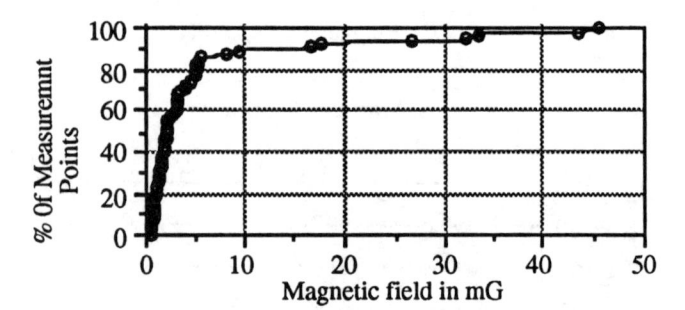

a. With Dc.

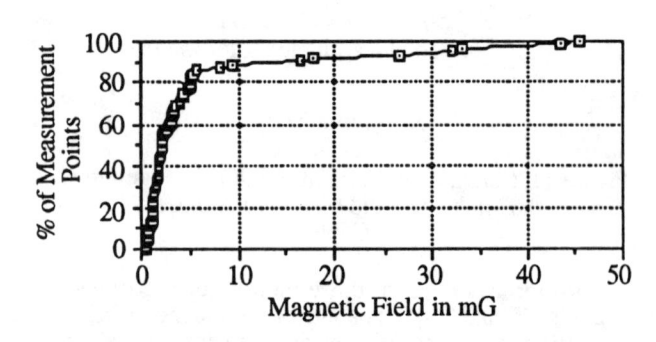

b. Without Dc.

Figure 13. Cumulative Distribution of Magnetic Field With and Without Dc Around the Vehicles.

Figure 14 shows the cumulative distribution of the THD of the magnetic field measured around the vehicles. The THD of the ac charger current (input current to the charger of a vehicle) and the THD of the magnetic field around the vehicle after the dc component was removed were compared, to discover if a correlation existed between the two. The average value of the THD of the magnetic field measured around the 30 electric vehicles is 63.35 % when the dc component is eliminated, about 13 % higher than the average value of the THD of ac charger currents. The difference is due to the fact that the magnetic field measured around electric vehicles is generated not only by the ac current of that charger, but the ac currents from nearby charger systems as well.

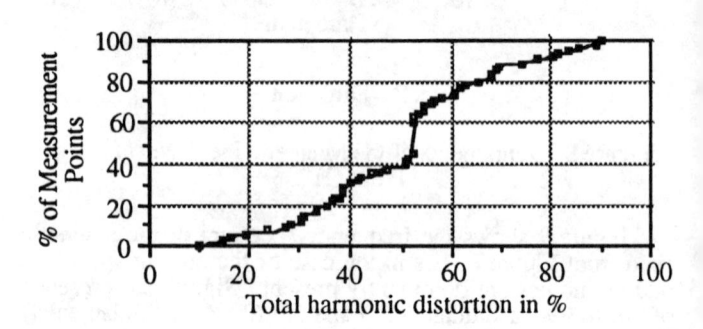

Figure 14. Cumulative Distribution of THD of Magnetic Field Around the Vehicles.

Comparison of the THD of the vehicles shows that the lowest THD (26.9 %) of the magnetic field was measured around vehicle number 8. The charger for vehicle number 8 is rated for 110 V ac input and 90 V, 10 A dc output, but this does not necessarily mean this particular vehicle generates a magnetic field with the least harmonics. Since battery chargers of other EVs are located nearby, there could be a cancellation of the high frequency components of the fields resulting in a lower THD.

At some points around the vehicle, the generated magnetic field is almost pure dc (point EV1 - vehicle 40, points EV1 and EV2 - vehicle 88 and point EV3 - vehicle 11, a pick-up). Superimposed on the dc magnetic field is a small ac component. Figure 15 illustrates this for the X direction of the magnetic field waveform measured at point EV1 - vehicle 40. The charger for vehicle number 40 is rated for 220 V ac input and 30 A dc output. The charger for vehicle number 88 is rated for 20 A ac input and 20 A dc output. Charger details for vehicle number 11 were not available.

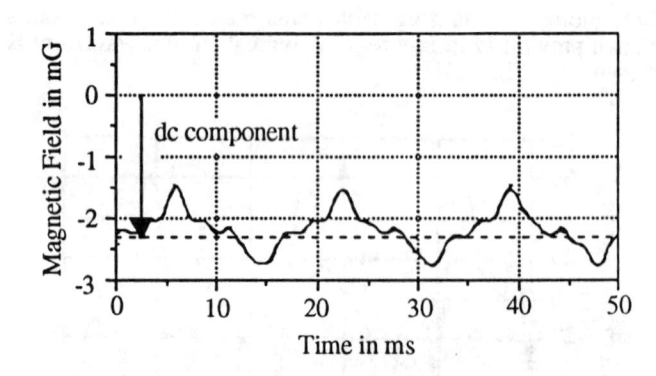

Figure 15. Dc Magnetic Field With Superimposed Ac Component.

The 60 Hz ac magnetic field with a peak value of 0.7 mG is superimposed on the 2.3 mG dc magnetic field. The dc component of this field is generated by the dc current from the battery charger to the batteries in the electric vehicle and the ac component of magnetic field is generated by the 60 Hz distorted periodic ac input current to the charger. Waveforms similar to those plotted for the trailer show the X, Y, and Z components of the magnetic field around the EVs are distorted.

DISCUSSION OF RESULTS

Data were collected on the magnetic field generated by EVs at the 1993 APS Solar and Electric 500. The field was measured at three locations: inside and around the charging trailer, along the charging cords, and around the EVs. The results, problems encountered, and suggestions for improvements are discussed below:

No dominant frequency components higher than 420 Hz were found. All chargers tested generated distorted periodic waveform currents. The 60 Hz frequency component was dominant in each of them. The average value of the THD of these currents is 50.1 %. The other major source of the magnetic field is the dc current from the chargers to the batteries. Since the dc current magnitude was higher than the ac current magnitude, the dc field is higher than ac field The dc field, though large in magnitude, was only about 10 % of the earth's field.

The magnetic field measured inside and outside the trailer had a very small presence of dc component except at location CT5, which was close to the EVs that were being charged. Thus, the field was mainly due to the ac currents. The maximum value of the magnetic field was measured near the circuit breaker which was located inside the trailer and had a magnitude of 64.37 mG.

The field generated by the cables bunched together on the ground was due to both ac and dc currents. The field was less than 20 mG for 80 % of the measurement points. The field magnitude reduced away from the receptacles because the charging cables were laid in a single row, bundled, and run on the surface between two rows of vehicles. The field was highest, therefore, near the receptacle panels (where all the cables originated) and decreased away from the panels. Instead of having one receptacle panel from which all charging cables were run to the EVs, 4-5 receptacle panels could be provided. This would result in a more uniform distribution of the field and a lower maximum value of the field near the charging cables and the EVs. People involved in the race were constantly moving around the cables, thereby exposing themselves. to the field The arrangement could be modified by running the cables underground, reducing the field above ground. The field above ground could be further reduced by shielding the cables.

The field round the vehicles is generated by the input ac current and the output dc currents of the charger. The dc field was less predominant when the charger was located on-board and more predominant when the charger was located off-board. The field was less than 20 mG for 90 % of the measurement points. Since the EVs were parked close to each other, it was difficult to isolate the various sources of the field and could give rise to higher field magnitudes at one point and lower field magnitudes at another point. Thus, it was difficult to compare the fields of the various vehicles. One way of overcoming these problems would be to increase the distance between adjacent vehicles.

The average value of the THD of the field was 63.5 % for the whole charging station. The average THD of the field without the dc component for inside and outside the trailer was 61.12 % which is close to the charging station average THD. The average THD of the field for the cables, without the dc component, was 51.5 %. This is close to the average THD of the ac charger currents. For the vehicles, without the dc component, the average THD of the field was 63.35 %. This is about 13.5 % higher than the ac charger currents and is due to the presence of other EVs nearby.

CONCLUSIONS

From the measurements and evaluations of the low frequency magnetic field data taken at the charging station at PIR the following conclusions can be drawn:

1. The magnetic field generated had both dc and ac components.
2. The highest value of the dc field measured was 50 mG and was measured near the charging cables. Though this value was comparable to the maximum value of the 60 Hz component, it was still only 10 % of the earth's magnetic field.
3. The charging station in a race environment generates low frequency magnetic fields. The strength of the magnetic field in the vicinity of the electric vehicles and battery chargers is in the range of 20 - 30 mG at a distance of 50 cm from them.
4. The maximum 60 Hz component of the magnetic field measured was 64.37 mG at a distance of 50 cm from the circuit breaker panels inside the trailer.
5. The generated magnetic field is a distorted periodic waveform with significant presence of odd harmonics. The average THD of magnetic field present in the charging station is 63.5 %.
6. Frequency spectrum analysis of the magnetic field in the charging station shows that only the first 7 harmonics have to be considered. Thus the frequency range of interest is 60 - 420 Hz.
7. The cables from the circuit breaker to the receptacle panel could be run underground, thereby reducing the magnitude of the field above ground.
8. The cables could also be shielded, reducing the field magnitude further.
9. Instead of providing just one receptacle panel from where all the charging cables to the EVs were run, 4-5 panels could be provided. These panels should be distributed within the charging station to avoid concentration of the field at just one point. This would give a more uniform distribution of the field and also the maximum field magnitude would be lesser.

REFERENCES

[1] G. G. Karady, Sh. H. Berisha, R. Hobbs, J. A. Demcko: "Electric Vehicle Magnetic Field Measurement", Paper Presented at SAE Future Transportation Technology Conference, August 1993, San Antonio, TX
[2] G. G. Karady, Sh. H. Berisha, T. Blake, R. Hobbs: "Power Quality at Electric Vehicle's Charging Station", To be Presented at the 1994 SAE International Congress and Exposition, Cobo Center, Detroit, Michigan, February 28 - March 3, 1994.
[3] G. G. Karady, Sh. H. Berisha, M. Muralidhar, J. A. Demcko, M. Samotyj: "Variable Speed Motor Drive Generated Magnetic Fields", To be Presented at the IEEE Power Engineering Society 1994 Winter Meeting, New York, New York, January 30 - February 3, 1994..
[4] EPRI Report "Magnetic Field Transient Resulting from Thyristor Controlled Loads", Dec. 1992.

940336

Specific Analysis on Electric Vehicle Performance Characteristics with the Aid of Optimization Techniques

P. Frantzeskakis, T. Krepec, and S. Sankar
Concordia Univ.

ABSTRACT

In this paper the effects of design parameters on the performance of an electric vehicle are presented. A detailed mathematical model was established using governing vehicle dynamics equations. Ideal energy storage systems were modelled with high order polynomial equations and represented graphically in the form of Ragonne curves. This was followed by the development of a simulation program which was utilized to optimize the design parameters, such as specific energy and mass of the storage system, electric motor operating voltage and electric drive final gear ratio. The effects these parameters had on the objective functions, namely range, acceleration, specific consumption, battery cycle life and cost were investigated. The outlined optimization process is presented in a manner which enables the designer to optimize electric or hybrid electric vehicles. Conclusions were drawn from the simulation and optimization results which would yield the best electric drive system for the vehicle in question.

INTRODUCTION

The demand for economical, and environmentally clean high performance vehicles has reached an all time high in today's society. This work is therefore, stimulated by the expected zero-emission vehicle (ZEV) regulations in the near future. Presently, electric vehicles can achieve this requirement, however, several problems such as, limited energy storage and cycle life capabilities of the battery system along with its cost, limit their use.

Research in testing and analysis of energy storage systems and electric drive units has been overwhelming in the past few years. However, a more scientific approach of selecting optimal design parameters is yet to be fully established.

To increase the performance of electric vehicles, proper selection of battery mass, operating voltage, energy storage systems and gear ratios is required. In order to design a complex engineering system whose performance depends on several conflicting objective criteria, multiobjective optimization techniques are utilized. In this paper, to properly understand the impact of these conflicting design parameters on the performance of electric vehicle systems, a graphical optimization method is performed. A detailed multiparameter optimization using nonlinear programming techniques is also carried out and the results are presented.

ELECTRIC DRIVE SYSTEM

The proposed electric drive system incorporates two electric motors installed in a Ford Escort Wagon. The first motor is a Solectria BRLS16 permanent magnet brushless dc motor and the second is an Advance DC series wound motor, model 203-06-401. Since, these motors function on different electromotive principles, synchronization is necessary. This is achieved by modifying the deadband region of the permanent magnet motor through the use of electric circuitry. Another aspect is the selection of the maximum power region of each motor. The Solectria BRLS16 motor provides peak power (22 kW) at a speed of 4500 rpm, whereas the peak power of Advance DC (33 kW) occurs at a speed of 4000 rpm. (See Fig. 1) The objective of such a design is to provide better acceleration, in comparison to a vehicle designed with two Solectria motors. During highway driving condition a control system disables the series wound motor, therefore the Solectria motor provides the entire power required to maintain the vehicle at the desired speed. This enables the Solectria to take up the full load, mapping an efficiency as high as 93%.

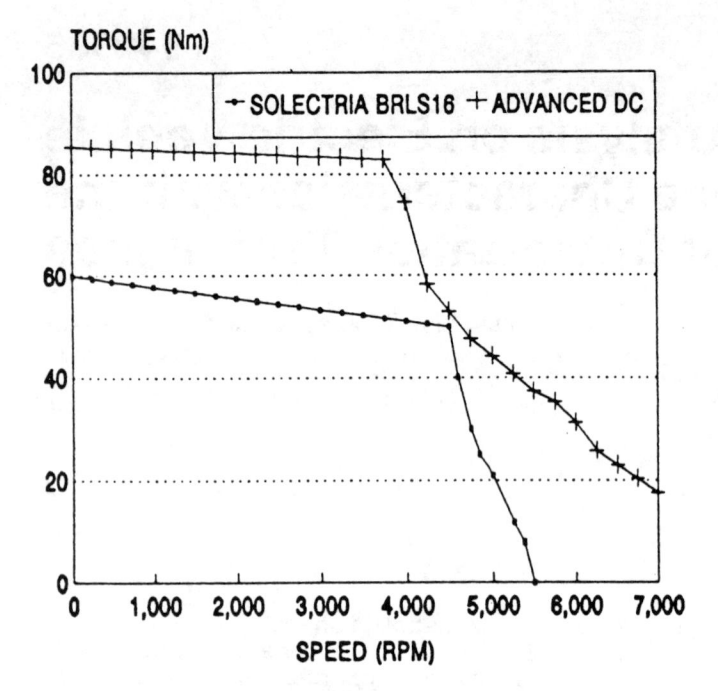

Fig. 1 Torque-RPM Curves, Solectria 144V and Advance DC 120V [1] [2]

The combined electric drive system is very attractive for it minimizes the cost of the electric drive by incorporating the series wound motor and controller as an acceleration unit, which is three and a half times less expensive than the permanent magnet motor and controller. The series wound motor provides sufficiently greater power; on the other hand, the regenerative braking features of the permanent magnet motor provide up to 15% of energy recuperation, during city driving. A chain drive concept enables both motors to bring power to the main shaft which is coupled to a 5 speed manual transaxle. (See Fig. 2)

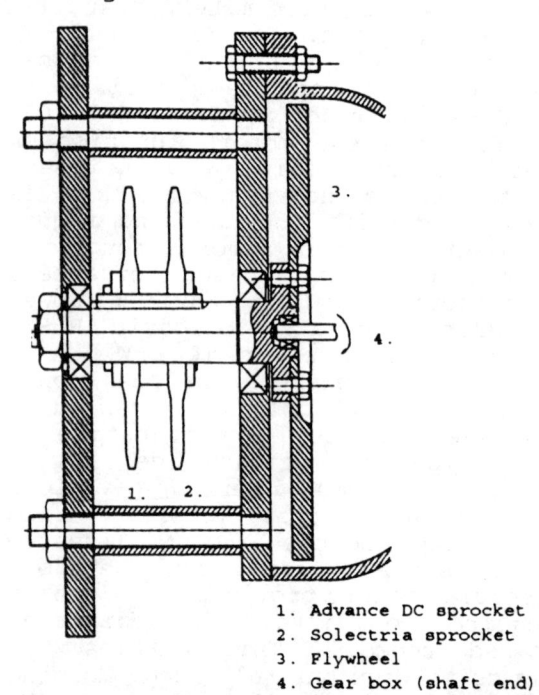

1. Advance DC sprocket
2. Solectria sprocket
3. Flywheel
4. Gear box (shaft end)

Fig. 2 Shaft Extension Assembly [3]

DESIGN CRITERIA

The objectives which were selected to analyze the design parameters are as follows:

- Time required to accelerate the vehicle up to a speed of 72 km/h,
- Range of the vehicle (full charge basis) travelling at 72 km/h on a flat surface,
- Energy capacity of the battery pack,
- Cost of the battery pack,
- Anticipated cycle life at 80% D.O.D. which is represented in terms of total obtainable range from the energy storage system during its entire life.

The design parameters of interest in this study are:

- Battery pack mass,
- Electric drive final gear ratio,
- Nominal operating voltage of the electric drive system,
- Specific energy of the storage system.

Design parameters such as tire rolling resistance, coefficient of drag and frontal area were excluded from the optimization study, since their effect on the vehicle performance is well known.

A maximum weight restriction of 425 kg was placed on the energy storage system, due to the GVW rating of the vehicle. The vehicle weight without any batteries was 1308 Kg. The coefficient of drag, frontal area and tire rolling resistance used in the analysis was $c_d=0.36$, $A_f=1.976$ m^2 and, $\mu_{const}=0.0085$, respectively.

MATHEMATICAL MODELLING

In order to perform a multiobjective optimization, appropriate mathematical models were developed for each objective function.

The mathematical model used to calculate acceleration and actual loads acting on the vehicle is outlined below. The three fundamental resistive forces acting on the vehicle are given below.

Aerodynamic drag force:
$$F_a = \frac{\rho C_d A_f}{2} (V_v \pm V_{wind})^2 \quad (1)$$

Rolling resistive force:
$$F_r = \mu_r mg\cos(\theta) \quad (2)$$

where
$$\mu_r = \mu_{const} + (0.0015 + \frac{V_v}{27.77}) \quad (3)$$

Climbing resistive force:
$$F_c = mg\sin(\theta) \quad (4)$$

Total resistive force:
$$F_{resistive} = F_a + F_r + F_c \quad (5)$$

Force developed at the wheels:
$$F_{dw} = (F_{solectria}\eta_{es} + F_{advance\,dc}\eta_{ea})\eta_m \quad (6)$$

Tractive effort:
$$F_{tractive} = F_{dw} - F_{resistive} \quad (7)$$

The maximum tractive force which can be developed is limited by the road adhesion, and is calculated by the following expression [4]. (See Fig. 3)

$$F_{dw}\,max = \frac{\frac{a\mu_a W}{L}}{1 + \frac{\mu_a h_{CG}}{L}} \quad (8)$$

By substituting, the following parameters:
· wheelbase length L= 2.541m
· center gravity height (h_{CG} = 0.508m),
· coefficient of road adhesion (μ_a = 0.85),
· and, a=0.49L under the assumption of a front/rear weight distribution of 51/49%.
The limitation of tractive effort, due to the road adhesion is obtained by simplifying equation (8):

$$F_{dw}\,max = 0.384\,mg$$

The effective mass of the vehicle is defined by the following relationship:

$$m_{eff} = m + \frac{J_{eff}}{r^2} \quad (9)$$

Where the effective inertia (J_{eff}) is calculated from the following expression:

$$J_{eff} = (J_a + J_s + J_{ed})(N_6 \cdot N_j \cdot N_s)^2 + (J_{gj} + J_{fw})(N_j \cdot N_6)^2$$
$$+ J_6(N_6)^2 + J_w \quad (10)$$

where:

N_s: electric drive final gear ratio
N_j: transaxle gear ratio j = 1 to 5
N_6: final drive ratio
J_{ed}: inertia of N_s
J_a: inertia of Advance DC motor
J_s: inertia of Solectria motor
J_w: inertia of wheels
J_{fw}: inertia of flywheel
J_6: inertia of N_6
J_{gj}: inertia of N_j

At this point, it is necessary to explain that the final drive ratio of the vehicle will be modified by the electric drive gear ratio, which will be determined by the optimization process.

The first objective function, acceleration time of the vehicle can, therefore, be calculated by the following expression:

$$t = m_{eff}\int_{V1}^{V2}\frac{dV}{F_{tractive}} \quad (11)$$

The cost and cycle life data which will be used in this optimization process, are estimated values, which have been obtain from testing and various other sources [5], [6], [7], [8], and [9]. A voltage to mass ratio was also calculated for all energy systems. This would provide an indication of the operating voltage which the energy system in question could obtain for a specified battery mass. (See Table 1)

The power required to maintain the vehicle at a constant speed of 72 km/h is calculated as follows:

$$Power = \frac{V_v F_{resistive}}{\eta_{es}\eta_m} \quad (12)$$

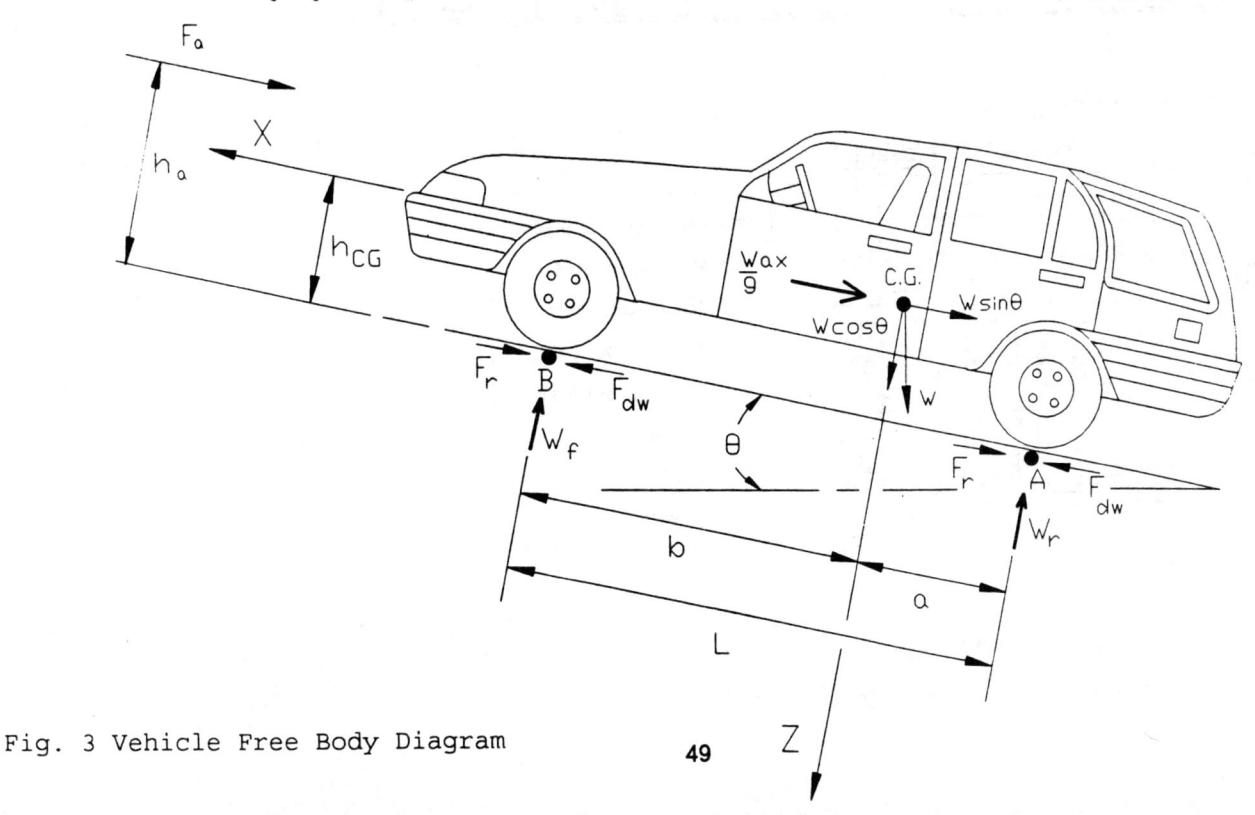

Fig. 3 Vehicle Free Body Diagram

The ideal characteristics of energy storage systems in Table 1 are represented by higher order polynomial which are thereafter used in the simulation program. The polynominal correlates energy density (x_i) as a function of power density (y); the subscript refers to the battery system. Therefore,

$$y = \frac{Power}{M_B} \quad (13)$$

$$x_i = a \cdot y^5 + b \cdot y^4 + c \cdot y^3 + d \cdot y^2 + e \cdot y + f \quad (14)$$

The Ragonne curves of ideal battery systems are shown in Figs. 4, 5 and 6.

The range which the vehicle can reach with a full battery charge can therefore be obtained from the following two equations:

$$hours = \frac{x_i}{y} \quad (15)$$

$$Range = V_v \cdot hours \quad (16)$$

Whereas, the total range the vehicle can travel over the entire cycle life of the energy system is as follows:

$$T_{range} = Range * Cycles \quad (17)$$

This is under the assumption of 72 km/h

Table 1: Cycle life and cost of energy storage systems

Battery Number x_i	Storage System	Specific Energy (Wh/kg) C/3	Cycle Life 80% D.O.D.	Estimated Cost ($/kWh)	Ideal Voltage mass ratio
1	Pb-Acid idle min.	27	450	80	0.71
2	Pb-Acid ideal max.	33	600	120	0.42
3	NiCd ideal min.	35	2000	1000	0.69
4	NiMH ideal min.	50	500	800	0.48
5	NiMH ideal max.	51	500	800	0.48
6	NiFe ideal	51	1000	800	0.23
7	NiCd intermed.	55	2000	1000	0.26
8	ZnBr ideal	56	500	500	0.50
9	NiCd ideal max.	64	2000	1000	0.26
10	NiZn ideal min.	72.8	600	350	0.27
11	NiZn ideal max.	79.3	600	350	0.27
12	NaS ideal min.	79	1000	650	0.35
13	NaS ideal max.	81	1000	650	0.36
14	AgZn ideal min.	117	100	1500	0.50
15	AgZn ideal max.	139	100	1500	0.50
16	Zn-Air ideal min.	144	150	1500	0.26
17	Zn-Air ideal max.	161	150	1500	0.26

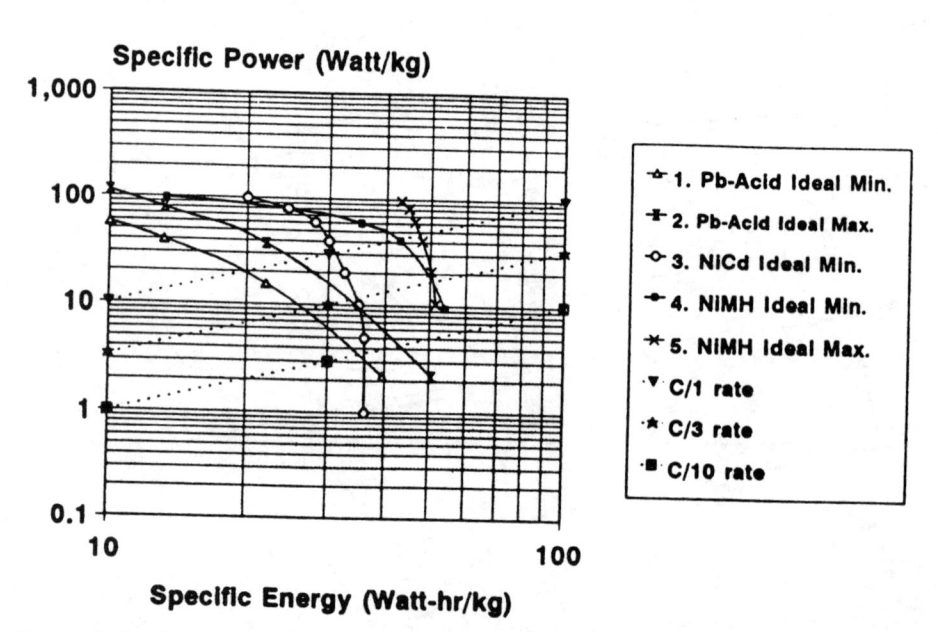

Fig. 4 Ragonne Curve - Ideal Energy Systems 1-5

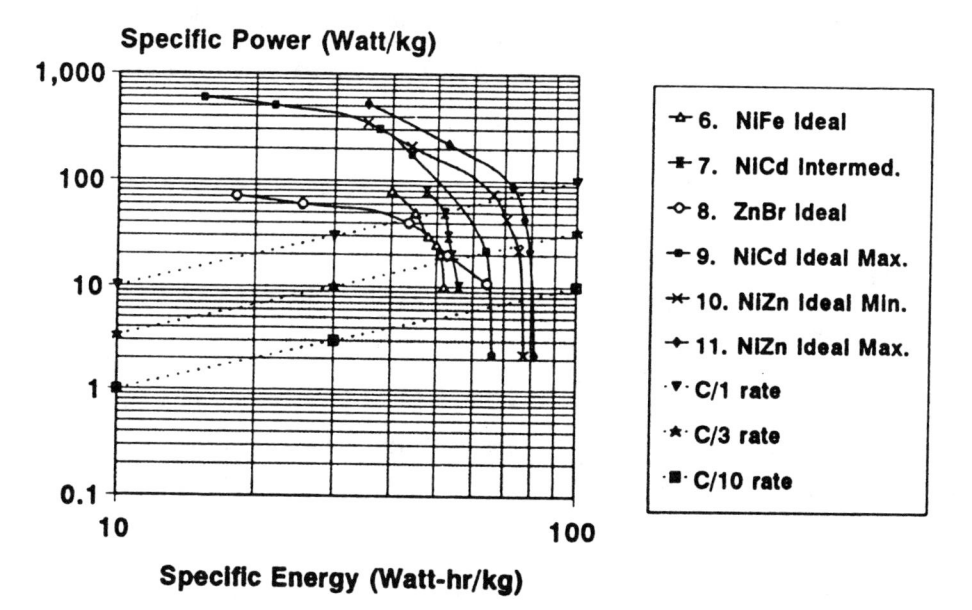

Fig. 5 Ragonne Curve - Ideal Energy Systems 6-11

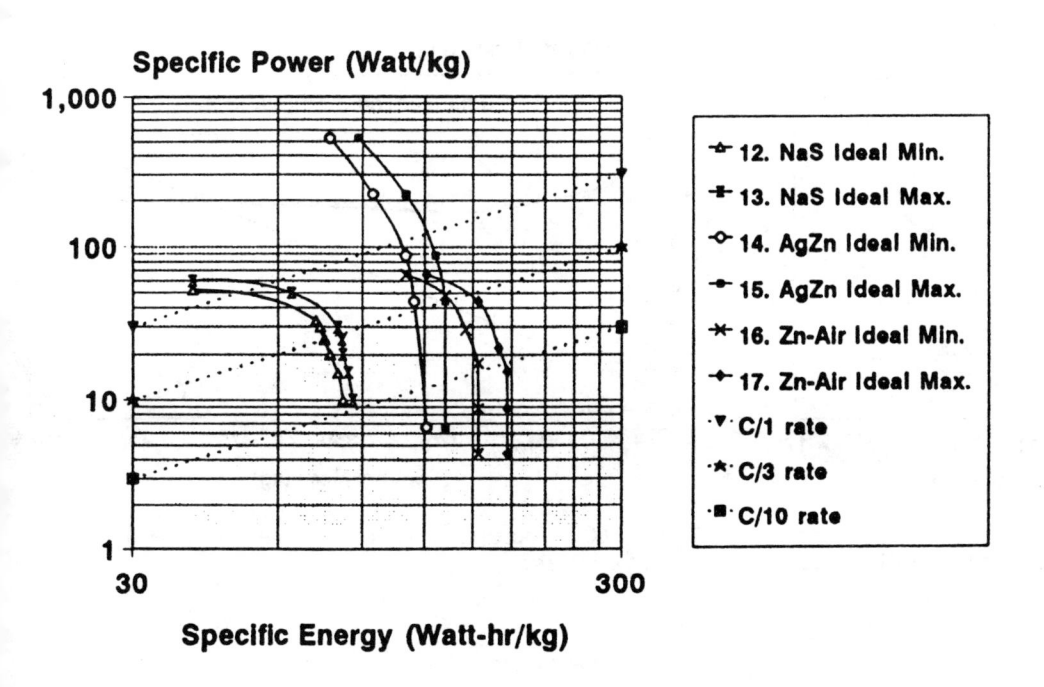

Fig. 6 Ragonne Curve - Ideal Energy Systems 12-17

constant speed, on a flat surface. The energy available in the battery system is then calculated as follows:

$$kWh_B = \frac{X_i \cdot M_B}{1000} \quad (18)$$

The cost of each energy system is calculated based on the C/3 specific energy rate and cost per kWh outlined in Table 1.

$$Cost = \frac{S_E(C/3)}{1000} M_B \cdot \frac{\$}{kWh} \quad (19)$$

DESIGN CONSTRAINTS

The lower and upper limits on design parameters are the constraints for the optimization process. These are represented below along with nominal values of design parameters.

$$275\,kg \leq M_B \leq 425\,kg \; ; M_{BN} = 350\,kg \quad (20)$$

$$1.0 \leq N_S \leq 3.1 \; ; \; N_{SN} = 2.04 \quad (21)$$

$$96 \, Volts \leq V_{OS} \leq 168 \, Volts; \; V_{OSN} = 132 \, Volts \quad (22)$$

$$27 \, \frac{Wh}{kg} \leq S_E \leq 161 \, \frac{Wh}{kg} \quad S_{EN} = 33 \, \frac{Wh}{kg} \quad (23)$$

GRAPHICAL OPTIMIZATION

In order to understand the engineering system which is being analyzed, a graphical optimization procedure was initially performed. By varying the design parameters within the constraints formulated previously, their impact on the objective functions can be investigated.

Each objective function is obtained by varying one design parameter, while holding the remaining nominal values constant. The first design parameter which was varied was energy density. The results are represented graphically in Figs. 7, 8, and 9. It can be clearly be seen that the only energy systems which have a fast acceleration time and relatively low cost, as energy density is varied, are 1,2,3,4,5,8,12, and 13.

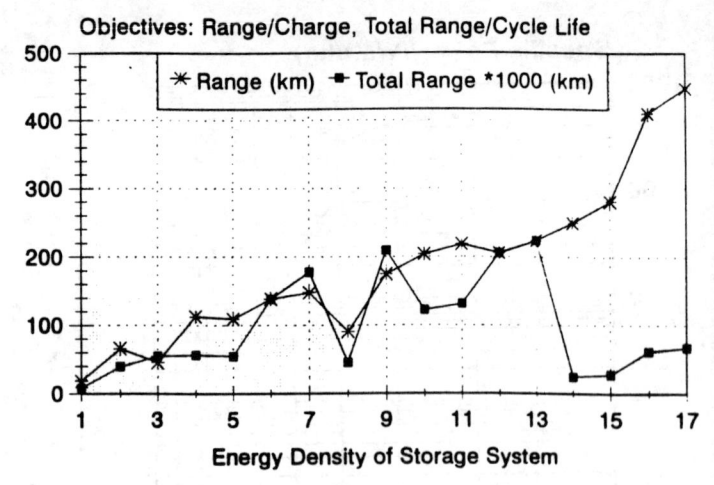

Fig. 9 Range and Total Range vs. Energy Density

As battery mass is varied, the effects of all objectives are clearly represented in the following figures. (See Figs. 10, 11, and 12)

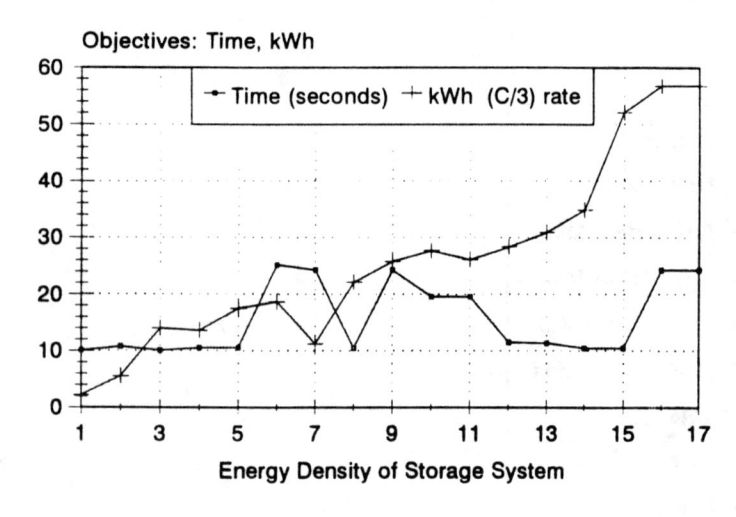

Fig. 7 Acceleration Time and kWh vs. Energy Density

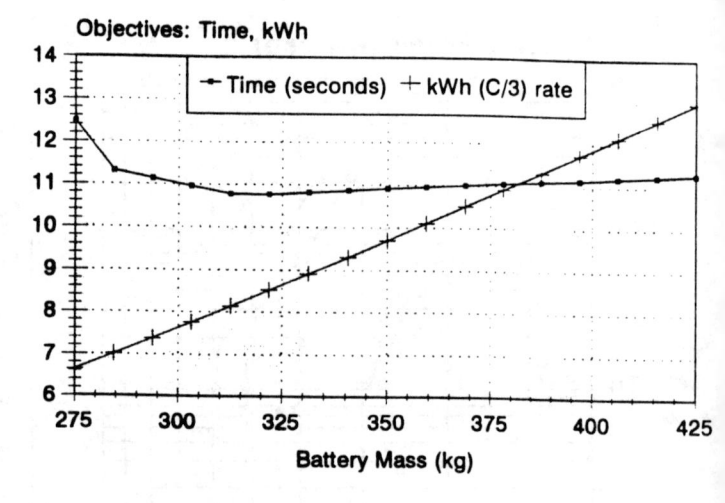

Fig. 10 Acceleration Time and kWh vs. Battery Mass

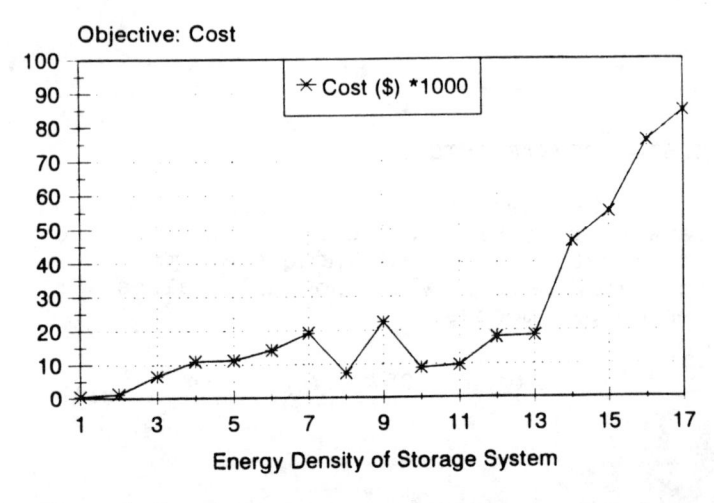

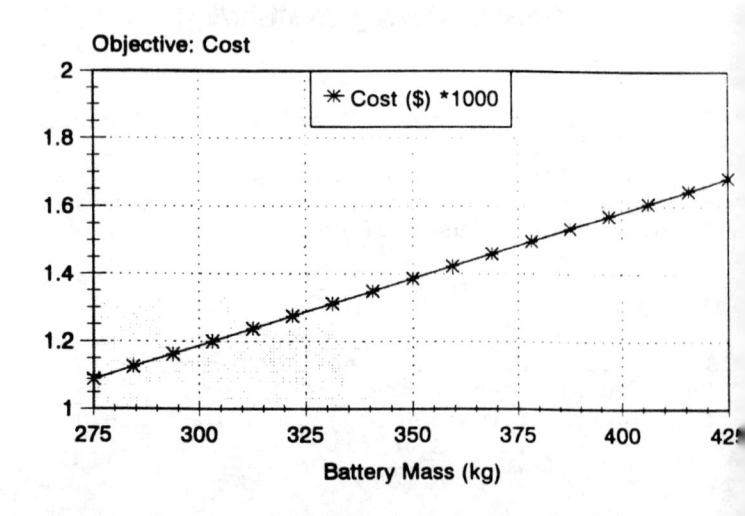

Fig. 8 Cost vs. Energy Density

Fig. 11 Cost vs. Battery Mass

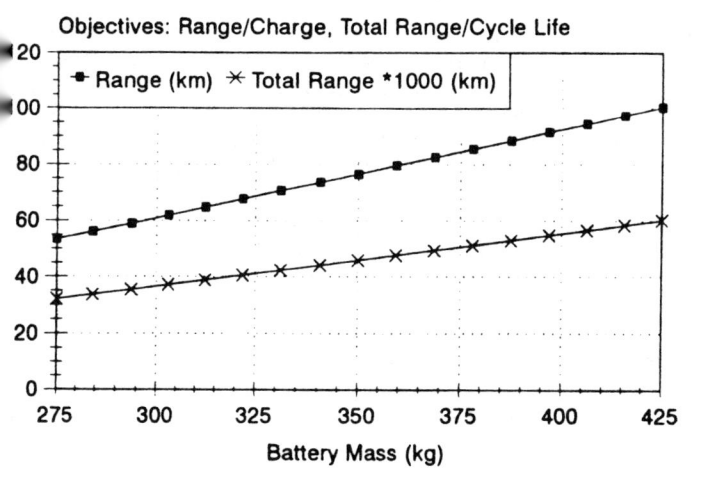

Fig. 12 Range and Total Range vs. Battery Mass

The electric drive final gear ratio has one distinct effect on the acceleration time, producing a global minimum within the feasible region at 2.05. (See Fig. 13)

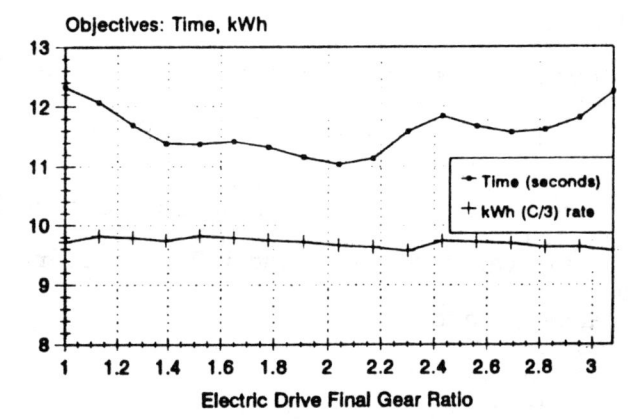

Fig. 13 Acceleration Time and kWh vs. Electric Drive Final Gear Ratio

As voltage is varied a saturation region occurs, which is created by the voltage to mass ratio outlined in Table 1. In order to obey the pre-specified nominal value of battery mass, the voltage effect is limited to 144 volts. (See Figs. 14 & 15)

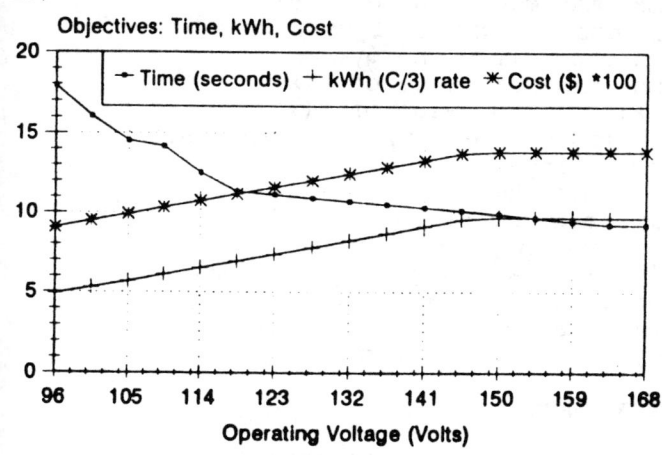

Fig. 14 Acceleration Time, kWh and Cost vs Operating Voltage

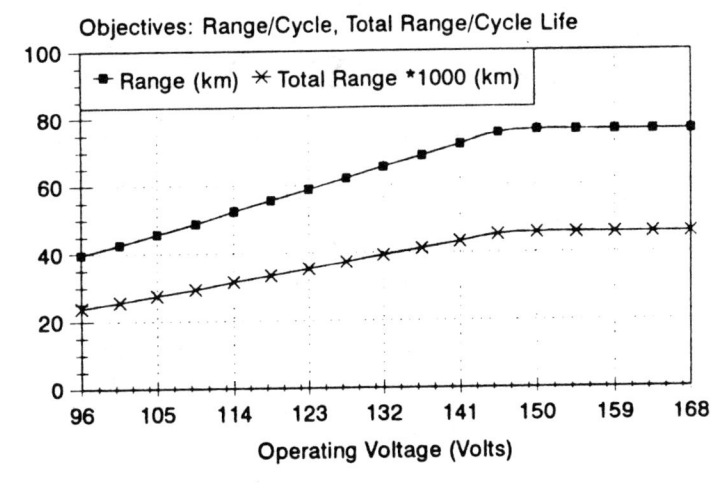

Fig. 15 Range and Total Range vs. Operating Voltage

With this complex optimization procedure, the graphical method of optimization becomes very time consuming because the nominal value selected for the energy storage system does not represent the general trend of the other system, due to the fact that they are nonlinear fuctions. However, a solution can be obtained at the expense of time by progressively changing the nominal value of the energy storage system and repeating the above procedure.

MULTIOBJECTIVE OPTIMIZATION

Alternatively, a multiobjective function was constructed in order to solve this optimization problem. The normalized multiobjective function is formulated as

$$F(x) = [\frac{T(x)-Tmn}{Tmx-Tmn}]^2 + [\frac{C(x)-Cmn}{Cmx-Cmn}]^2 + [\frac{K(x)-Kmn}{Kmx-Kmn}]^2 + [\frac{R(x)-Rmn}{Rmx-Rmn}]^2 + [\frac{TR(x)-TRmn}{TRmx-TRmn}]^2 \quad (24)$$

By choosing five weighting values w_1, w_2, w_3, w_4 and w_5 and applying them in equation (24), the global criterion function can be rewritten as:

$$F(x) = w_1[\frac{T(x)-Tmn}{Tmx-Tmn}]^2 + w_2[\frac{C(x)-Cmn}{Cmx-Cmn}]^2 + w_3[\frac{K(x)-Kmn}{Kmx-Kmn}]^2 + w_4[\frac{R(x)-Rmn}{Rmx-Rmn}]^2 + w_5[\frac{TR(x)-TRmn}{TRmx-TRmn}]^2 \quad (25)$$

The objective function in equation (25) together with the design constraints in equation (20) to (23) are solved using an interior penalty function method.

The pseudo-objective function is

defined as:

$$\phi(X) = F(X) - p_p \sum_{i=1}^{m} \frac{1}{g_i(X)} \quad (26)$$

where $g_i(X)$ represents the inequality constraints outlined previously and p_p is the penalty parameter whose initial value is selected as:

$$p_p = \frac{F(X_1)}{-\sum_{i=1}^{m} \frac{1}{g_i(X_1)}} \quad (27)$$

and thereafter decremented. Minimization of the pseudo function is obtained by using the Hooke and Jeeves method [10] and [11].

The weighting factors (w_i's): 0.3, 0.5, 0.1, 0.05, and 0.05 were used on the global objective function. In return, the pseudo-objective function obtained the following design parameters:

- Battery mass: M_B = 400 kg,
- Electric drive final gear ratio: N_s = 2.05,
- Nominal operating voltage: V_{os} = 168 Volts,
- Specific energy density of the storage system: S_E = 33 Wh/Kg.

The proposed weighting scheme places the majority of the emphasis on acceleration time and cost of the energy system. These weighting factors can be changed in order to suit the designers needs.

Performance Analysis Based on Calculated Optimal Design Parameters

The calculated optimal design parameters were used as inputs for the simulation previously outlined. The anticipated performance of the vehicle is as follows:

- 9.61 seconds (0 to 72 km/h),
- 1600 $ battery pack cost,
- 11.9 kWh battery pack energy,
- 93 km range (full charge basis) travelling at 72 km/h on a flat surface,
- 55 900 km obtainable range from the energy storage system during its entire life.

CONCLUSIONS

The effects of design parameters on the performance of an electric vehicle were presented. A detailed mathematical model was established using governing vehicle dynamics equations. Ideal energy storage systems were modelled with high order polynomial equations and represented graphically in the form of Ragonne curves. This was followed by the development of a simulation program which was utilized to optimize the design parameters, such as specific energy and mass of the storage system, electric motor operating voltage and electric drive final gear ratio. The effects these parameters had on the objective functions, namely range, acceleration, specific consumption, battery cycle life and cost were investigated. A detailed multiparameter optimization using nonlinear programming techniques and a graphical optimization was carried out. The simulation and optimization results for the proposed weighting scheme on the global criterion function and design constraints, which would yield the optimal electric drive system for the vehicle in question, were also presented.

ACKNOWLEDGEMENTS

The authors express their gratitude to FCAR, NSERC, Energy, Mines, and Resource Canada, Argonne National Laboratory, Ford Motor Company and Society of Automotive Engineers.

NOMENCLATURE

V_v: vehicle velocity
V_{wind}: wind velocity
ρ: density of air
n_m: mechanical efficiency
n_{es}: Advance DC electric motor efficiency as a function of rpm
n_{ea}: Solectria motor efficiency as a function of rpm
$F(x)$: multiobjective function
g_i: inequality constraint on design parameters
m: total vehicle mass including battery mass
M_B: battery mass
M_{BN}: nominal M_B
V_{os}: operating voltage
V_{osN}: nominal V_{os}
S_E: specific energy of storage system
S_{EN}: nominal S_E
N_s: electric drive final drive ratio
N_{sN}: nominal N_s
$T(x)$: acceleration time
Tmn: minimum value of $T(x)$
Tmx: maximum value of $T(x)$
w_1: acceleration weighting factor
$C(x)$: cost
Cmn: minimum value of $C(x)$
Cmx: maximum value of $C(x)$
w_2: cost weighting factor
$K(x)$: specific energy
Kmn: minimum value of $K(x)$
Kmx: maximum value of $K(x)$
w_3: specific energy weighting factor
$R(x)$: range
Rmn: minimum value of $R(x)$
Rmx: maximum value of $R(x)$
w_4: range weighting value
$TR(x)$: total range
TRmn: minimum value of $TR(x)$
TRmx: maximum value of $TR(x)$
w_5: total range weighting factor

REFERENCES

1. Solectria Electronic Division, Arlington, MA, Pamphlet.
2. Advance D.C. Motors Inc., Syracuse,

New York, Pamphlet.
3. Dionatos, D., Frantzeskakis, P., Kekedjian, H., Nikopoulos, A. and J., Theofanopoulos, " The Concordia University Hybrid Electric Conversion," SAE SP-980.
4. Gillespie, D.T., "Fundamentals of Vehicle Dynamics," Society of Automotive Engineers, Inc., Warrendale, PA, 1992.
5. DeLuca, W.H., Gillie, K.R., Kulaga, J.E., Smaga, J.A., Tummillo, A.F. and Webster, C.E. " Performance and Life Evaluation of Advanced Battery Technologies for Electric Vehicle Applications," SAE Paper No. 911634.
6. Dickinson, B.E., Swan, D.H. and Lalk T.R., "Comparison of Advanced Battery Technologies for Electric Vehicles," SAE Paper No. 931789.
7. Reisner, D. and Eisenberg, M. " A New High Energy Stabilized Nickel-Zinc Rechargeable Battery System for SLI and EV Applications," SAE Paper No. 890786.
8. Ragone, D.V. " Review of Battery Systems for Electrically Powered Vehicles," SAE Paper No. 680453.
9. Kucera, G., Plust, H.G. and Schneider, C. "Nickel-Zinc Storage Batteries as Energy Sources for Electric Vehicles," SAE No. 750147.
10. Rao, S.S., "Optimization - Theory and Practice," John Wiley & Sons, New York, 1984.
11. Tummala, M., Krepec, T. and Ahmed, A.K.W. "Simulation, Testing and Optimization of Natural Gas On-Board Storage System for Automotive Applications," SAE Paper No. 931820.
12. Danczyk, L.G., Scheffler, R.L. and Hobbs, R.S. " A High Performance Zinc-Air Powered Electric Vehicle," SAE Paper No. 911633.
13. Henrikesen, G.L.,Patil, P.G., Ratner, E.Z. and Warde, C.J. " Assessment of EV Batteries and Application to R&D Planning," SAE Paper No. 890781.
14. Shemmans, J.M., Sedgwick D., Pekarsky, A. " NaS Batteries for Electric Vehicles," SAE Paper No.900136.
15. Karl V. Kordesch, "Batteries", Marcel Dekker Inc., New York, 1977.
16. Tuck, C.D.S., "Modern Battery Technology," Ellis Horwood Ltd., West Sussex, England, 1986.
17. Papalambros, P.Y. and Wilde, D.J., "Principles of Optimal Design," McGraw-Hill Book Company, New York, 1978.
18. Rao, S.S., "Design of Vibration Isolation Systems Using Multiobjective Optimization Techniques," ASME Paper No. 84-DET-60.

940337

The Development and Performance of the AMPhibian Hybrid Electric Vehicle

Gregory W. Davis, Gary L. Hodges, and Frank C. Madeka
United States Naval Academy

ABSTRACT

The design specifications and the results of the performance and emissions testing are reported for a series Hybrid Electric Vehicle(HEV) which was developed by a team of midshipmen and faculty at the United States Naval Academy. A 5-door Ford Escort Wagon with a manual transmission was converted to a series drive hybrid electric vehicle. The propulsion system is based on a DC motor which is coupled to the existing transmission. Lead-acid batteries are used to store the electrical energy. The auxiliary power unit(APU) consists of a small gasoline engine connected to a generator. All components are based upon existing commercial technology.

INTRODUCTION

A series Hybrid Electric Vehicle(HEV) has been developed by a team of midshipmen and faculty at the United States Naval Academy for use in the Hybrid Electric Vehicle Challenge which took place during June of 1993. This competition, involving thirty universities from North America, was jointly sponsored by Ford Motor Company, the SAE International, and the U. S. Department of Energy. A 5-door Ford Escort Wagon with a manual transmission has been converted to a series drive hybrid electric vehicle. The propulsion system is based on a DC motor which is coupled to the existing transmission. Lead-acid batteries are used to store the electrical energy. The auxiliary power unit(APU) consists of a small gasoline engine connected to a generator. The AMPhibian is designed to be an economically feasible HEV, for use in near term applications. To accomplish this, all components are based upon existing commercial technology. Further, this vehicle was designed to retain, to the greatest degree possible, the basic driving characteristics of a conventional gasoline powered vehicle.

DESIGN OBJECTIVES

The challenge involved many aspects including cost effectiveness, acceleration, range, safety, and emissions, which were incorporated into the vehicle design.

COST - Since the AMPhibian was designed to be economically feasible, minimizing cost was considered to be a major design goal. All design decisions were made only after the associated costs were analyzed. To help attain this goal, all components were based upon existing, available technology.

PERFORMANCE AND EMISSIONS - The major performance and emissions design goals for the AMPhibian include 1) the ability to travel 64 Km as a zero emissions vehicle(ZEV) using battery power alone, 2) operating in hybrid mode, the ability to travel 320 Km while meeting the transitional low emissions vehicle(TLEV) air pollution standards, 3) achieve a time of under 15 seconds when accelerating from 0 to 70 Kph, and 4) climb a minimum of a 15% grade. The vehicle was also to maintain driving characteristics as similar to that of conventional gasoline powered vehicles as possible(e.g. one brake pedal, shift gears normally, etc.).

RELIABILITY AND DURABILITY - The AMPhibian should have reliability and durability similar to that of a conventional gasoline powered vehicle. Using existing components not only helps to limit the costs, but also to help ensure reliable and durable operation of the vehicle.

SAFETY - Occupant safety was a prime concern. The frontal impact zone and original vehicle bumpers were maintained to provide sufficient collision protection. The original power-assisted braking system also remained intact to ensure proper braking. A fire suppression system was added to the vehicle and battery compartments, as well as to the engine bay to minimize the chances of injury and equipment damage. Due to the additional vehicle weight, the roof structure was augmented to provide additional protection in case of a vehicle roll-over. Finally, the competition rules required the use of a five point harness system for both the driver and passenger.

WEIGHT - One major disadvantage of electric vehicles has traditionally been the large weight due to the propulsion batteries required to provide the energy storage capability for extended range. An advantage of the HEV concept is to allow for less energy storage capability of the batteries by replacing some of these batteries with a small auxiliary power unit(APU) which provides the equivalent amount of energy with less weight. However, battery weight is still considered to be a major concern, requiring the team to consider all options for reducing vehicle weight. The AMPhibian was designed to weigh less than the gross vehicle weight rating(GVWR) of the 1992 Escort LX Wagon plus an additional 10%. This results in a maximum allowable vehicle mass of 1729 kg. Further, to maintain acceptable handling, the side-to-side bias must remain within 5% of neutral, and the front-to-rear bias must not drop below about 40%/60%.

PASSENGERS AND CARGO - The HEV carries one driver and one passenger, along with a volume of cargo(50 cm by 100 cm by 25 cm). The total combined weight of people and cargo is a minimum of 180 kg.

BATTERY CHARGING - The HEV charging system was designed to recharge the battery pack in six hours. This should reduce daytime charging demand on electrical utilities. Daytime charging, if necessary, could be accomplished using the APU. The charging system accepts either 110V or 220V, 60 Hz AC power.

STYLING - Vehicle styling changes were minimized to maintain continuity with existing vehicle designs. No external glass or body sheet metal was modified except to provide additional ventilation.

VEHICLE DESIGN

The relationship of the design goals was studied, and compromises were made to provide near optimal system design, given the severe budgetary and time constraints. This process resulted in the selection and design of the major vehicle components. The following discussion details the design decisions and vehicle specifications which are summarized in table 1.

POWERTRAIN - The AMPhibian is propelled using a series drive configuration. That is, the only component that is mechanically connected to the drive-train of the vehicle is the electric motor. This arrangement is depicted in fig. 1, a more detailed electrical schematic is shown in fig. 2. This arrangement was considered to be superior to the parallel drive arrangement, in which both the electric motor and the APU can propel the vehicle, for the following reasons. The series drive would require less structural change to install, and thus provide a lower cost. The parallel drive system would also require a more sophisticated control system to minimize driveability problems such as those associated with the transition from electric vehicle(EV) mode to hybrid electric vehicle(HEV) mode. This would, again, result in higher cost, and, possibly, reliability problems due to the added complexity.

The conversion to a series drive system required the removal of the standard Escort engine. Since the Escort has front-wheel drive, the standard engine is mounted transversely in a transaxle arrangement. Thus, the transaxle was left intact so that a new axle would not need to be designed. The electric motor was attached directly to the existing bell-housing and flywheel. This arrangement also allows full use of the existing transmission, thus allowing for variable gear ratios. This was considered an advantage since it would allow the electric motor to be operated closer to its preferred operating speed over varying vehicle speeds.

Prior vehicle testing and simulation indicated that the vehicle would require a power of approximately 9 kW in order to maintain a steady 80 Kph. Acceleration from a stand still to 72 Kph in less than 15 seconds would require a peak power of 32 KW(at approximately 35 Kph) for a short

Table 1. Summary of Components used in the U. S. Naval Academy's Hybrid Electric Vehicle.

Chassis:	'92 5-door Escort LX
Stock GVWR:	1572 kg
Converted GVWR:	1729 kg
Maximum Carrying Capacity(passengers and cargo):	180 kg
DC Motor:	General Electric model 5BT1346B50
Motor Controller:	Curtis PMC 1221B-1074
Batteries (propulsion):	10 arranged in series, 12VDC Trojan 5SH(P)
Bus Voltage:	120 VDC
APU Engine:	Briggs and Stratton Vanguard V-twin, 13.4 kW @3600 RPM, two cylinder
APU Alternator:	Fisher Technology, Inc., 13.5 kW, 150 Vpeak
Tires:	Goodyear Invicta GL P175/65R14, low rolling resistance
Estimated Vehicle Cost:	$26,000
Conversion Component Net Cost: (exc. safety items, credit for 1.9l engine)	$14,000

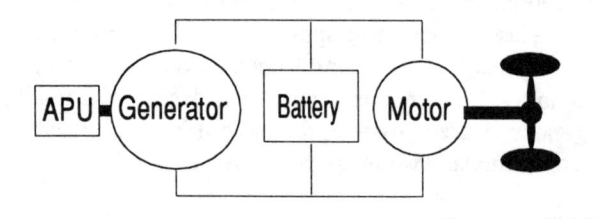

Fig. 1. Series Drive Diagram for the U. S. Naval Academy's Hybrid Electric Vehicle.

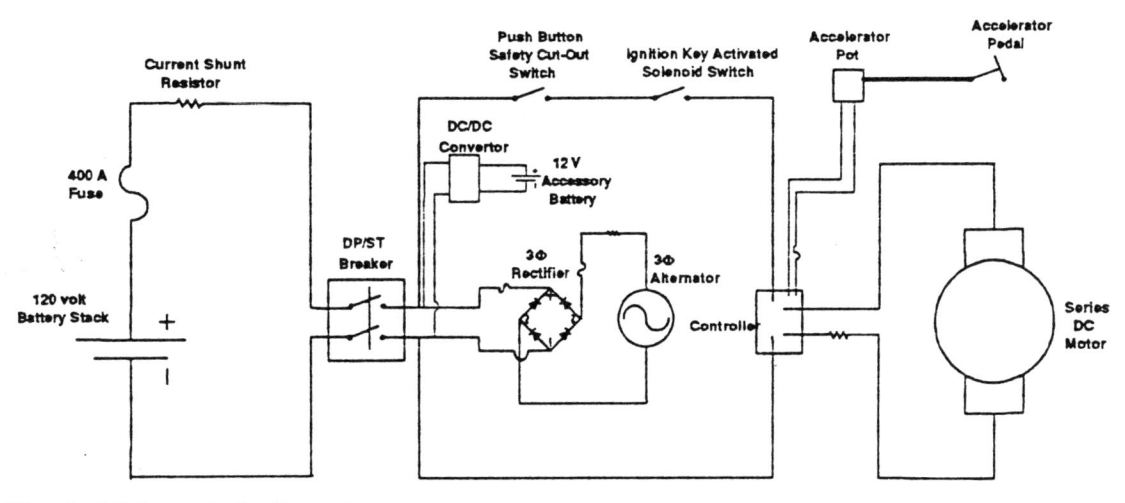

Fig. 2. Electrical Schematic for the U. S. Naval Academy's Hybrid Electric Vehicle.

duration. Motor controller cost and availability became the critical design factor for the selection of both the type of motor and the system operating voltage. The use of an AC motor was investigated due to its inherently higher power density compared to a DC system. However, it was rejected due to the cost, availability, size, and weight of the associated motor controller. A series connected, 15.2 kW(@ 90 VDC) DC motor was chosen instead since DC motor controllers are more widely available, less costly, and lighter in weight. The combination DC motor and controller weighs approximately 82 kg, the engine that was removed weighed 113 kg, thus resulting in a net weight savings of 31 kg. Although the steady state rating is less than the peak incurred during the acceleration, the motor can provide a peak power 2-3 times its steady state rating for short duration. A controller rated at 120 VDC(160 V peak) was chosen, thus this determined the system operating voltage.

BATTERY SELECTION - The AMPhibian has two battery power systems. One system is at 12V and one at 120V. The 12V system is used to power the 12V lighting and accessories. The 120V primary battery powers the prime mover and supplies power to recharge the 12V battery.

The battery selection was overwhelmingly driven by cost considerations. Secondary considerations included: 1) the HEV Challenge constraint of 400V or less battery stack voltage, 2) the motor controller rating of 120V, 3) the HEV Challenge constraint of no more than 20 kW-hr capacity at a 3 hr discharge rate, 4) the gross vehicle weight rating constraints and 5) practical considerations. In general, an inexpensive, small, lightweight battery having high specific power and high specific energy is desired for use in the AMPhibian. Additional considerations included the desire to maximize voltage thereby minimizing power losses due to the lower operating currents. Also, to help to maximize electrical energy storage capacity, and, therefore, ZEV capabilities, the battery ampacity rating should be maximized. Since the maximum rating for the motor controller is 120V, 120V was selected. An order-of-magnitude calculation of the costs of batteries having characteristics superior to those of conventional lead-acid batteries lead the design team to limit selection considerations to off-the-shelf lead-acid batteries. For example, Nickel-Iron batteries were found at a cost of $1800 per six volt battery or $36,000 for a 120V battery stack. Nickel-Cadmium were found at a cost of $964 per six volt battery or $19,280 for a 120V battery stack. Both estimates far exceeded AMPhibian budget constraints; therefore only lead-acid batteries were considered.

The task of battery selection was complicated due to the general lack of published, comprehensive, technical battery performance data covering an extensive number of battery models and manufacturers which had been verified by an independent source. This limited information resulted in the selection of the Trojan 5SH(P) battery. The Trojan 5SH(P) battery is a deep-cycle, wet-celled, 12V battery. The "L" type terminals were selected for this application. With the primary battery selected, the 12V system needed to be defined and selected.

Several approaches were considered to power the 12V system. This included the extremes of using the existing 12V system, as is, or converting all 12V components to 120V. Engineering judgment indicates the latter option is not practical. One approach for providing power to the 12V system was to utilize the output of one twelve volt battery from the 120V stack. This approach has the advantage of simplicity. One disadvantage of this approach is that, using the existing 12V components which are grounded to the chassis, means that the battery stack is no longer electrically isolated from the chassis and, thus, the chance of injury in the event of failure is increased. Another problem, is that, since the batteries are connected in series, if the battery used for the 12V system fails, the whole battery stack will become inoperable. The chance of battery failure can be reduced by inserting a higher amp-hr rated battery into the 120V stack to compensate for the added use. The disadvantage is that this local change to the series of batteries imparts an unknown on the primary battery stack performance (i.e., internal impedance and resistance). This lead to the decision to have two separate battery systems, a 120V primary system and separate 12V system.

Several options were considered for the 12V system. One option was to incorporate a single, independent high amp-hr rated battery required to provide several hours at a relatively high discharge rate (e.g., driving at night and in rain/use of head lights and wipers). This battery would then be recharged externally during refueling and/or recharging. This option was rejected due to the resulting high weight of the battery. A DC/DC converter, powered by the 120V stack, could be used to meet all of the 12V demand. However, this converter must meet the peak 12V load, which is estimated to be 210 amps during starting of the APU. This option was rejected due to the heavy weight and size of this converter. A DC/DC converter, sized to handle the sustained accessory loads under moderate to heavy use, was incorporated in parallel with a small 12V battery, sized to accommodate the APU starting loads. This design saves both space and weight. The estimated sustained load encountered during moderate to heavy accessory use is 20A. However, a 30 amp DC/DC converter was selected to accommodate the future addition of a climate control system for the passenger compartment. The APU starter requires a battery rated at 210 cranking amps. The ultra light Pulsar Racing Battery, offered by GNB Incorporated, was used since this battery weighed only 4.5 kg, or approximately 50% less than other conventional lead-acid batteries, and provides 220 cranking amps. AMPhibian's net 12V accessory system, occupies the same volume as the OEM 12V battery, but weighs approximately 9.5 kg less than the OEM battery alone.

AUXILIARY POWER UNIT - The design specifications for the auxiliary power unit (APU) were derived from the mechanical power necessary to achieve the 320 km desired range while maintaining highway speeds, and allowing for reasonable accelerating and coasting time periods with the batteries at 20% of full charge at the beginning of APU operation. Calculations based on these estimates of driving conditions (drag and rolling resistance) and drivetrain efficiency resulted in a minimum desired electric power availability of 10 kW. If the APU could deliver this power, the HEV would be able to sustain highway speeds for the full range, limited only by the amount of onboard fuel. However, this power capability alone would not allow for reasonable accelerations over this distance. Therefore, the APU must be capable of charging the batteries while at highway speeds so that if acceleration becomes necessary, the power may be drawn from both the batteries and the APU.

The total calculated electrical requirement resulted in a specification of 12.5 kW output from the APU. Estimating the overall efficiency of the APU to be 80%, the engine then must be capable of mechanically developing 15.6 kW.

With the design parameters determined, the selection of the actual components centered around availability of "shelf" items, size and space limitations, emissions and ultimately and most significantly the cost. Ideally an engine-generator set could be found meeting all the requirements. However, a review of the available market provided no likely candidates, particularly in terms of weight and space requirements. Therefore, the engine and generator were selected separately.

Based upon time constraints and availability, the design team limited the choice to a conventional, gasoline powered spark-ignited engine. Briggs & Stratton donated a 13.5 kW "Vanguard" series engine. Although this engine did not meet the expected power demand, cost considerations dictated its use. Speed is regulated by a governor to 3600 RPM and is adjustable. This engine weighs 40 kg, fits under the hood, and has a pull-cord starting mechanism. To meet TLEV emissions requirements the APU exhaust is connected to a catalytic converter which contains a ceramic monolith substrate. The outlet of the catalytic converter will then lead to the existing vehicle exhaust system. An electrically driven air pump was added to provide fresh air for the catalytic converter after light-off to ensure complete oxidation of unburned hydrocarbons and carbon monoxide.

To meet the electrical requirements, a number of alternatives involving both AC and DC generation were explored. To minimize space and weight, a custom built alternator was considered the best choice. Although this was the most expensive component of the vehicle, an off-the-shelf item was unavailable. A vendor was contracted to build a custom 13.5 kW, 150 V, 3 phase alternator. The voltage was selected based on providing 144 V DC from a three phase bridge rectifier, the maximum recommended charging voltage for a 120 V battery stack. However, this voltage is well above the rated voltage for the controller. Therefore, a 50 watt 130 V zener diode is used to limit the controller voltage while charging. Additional voltage control can be obtained by varying the speed of the APU. Beside meeting all electrical requirements, this alternator weighs only 4.6 kg, is 0.276 m in diameter and when mounted directly on the APU shaft, extends a mere 0.19 m from the engine block, making it the most feasible option.

The strategy for controlling the APU was developed for manual operation and will be incorporated into a digital control system with a goal of making driving the HEV as much like driving a conventional vehicle as possible. The system must sense battery voltage to determine battery condition and control the APU operation. It must also control ancillary functions such as insuring the battery box exhaust fan is on whenever charging and monitoring APU current to prevent overload.

A three position switch was mounted in the passenger compartment to start or stop the APU or place control in the strategy mode. The first two positions are self explanatory and will override the strategy mode, allowing the driver to make the decision to start or stop the APU while monitoring battery voltage, motor and APU current along with other pertinent parameters. However, while in the strategy mode, the system will be make decisions and carry out programmed actions, while signaling with status lights to keep the driver informed.

SAFETY - To enhance the roof structure, a Sports Car Club of America approved roll-bar was purchased and installed in the vehicle. Obviously, the structure of the roof itself in a production vehicle would be enhanced to meet the additional weight demands.

In the event of fire, a halon fire suppression system was

installed. This system can provide a significant suppression of an electrical fire.

The original power-assisted braking system was to remain intact to ensure proper braking. Since the engine was removed, the vacuum assist was disabled, therefore an electrically powered vacuum pump and reservoir were installed to replace this loss of vacuum.

Additional safety devices include a panic switch that can be used to disconnect the battery pack from the vehicle in the event of emergency. An in-line fuse and single-throw, double-pole circuit breaker were also installed to add additional redundancy for protection of the occupants and equipment. The circuit breaker also serves to isolate the batteries to allow for safe maintenance and testing of other components. Since a high voltage and current system is used, the chassis is not used as the ground as is the usual case in conventional vehicles. Both positive and negative cabling are used to minimize stray currents and voltages, and to isolate the system from the chassis. Additionally, circuit breakers, both 110V and 220V, have been installed between the on-board charger and the external power connections to ensure safe charger operation independent of the power source.

CHARGING SYSTEM - A on-board MOSFET battery charger was chosen for use in the vehicle. This charger is both lightweight and can accept either 110 V or 220 V, 60 Hz, AC power. To reduce weight, an isolation transformer is used as an off-board component. It was felt that future infrastructure could provide adequate isolation at the stationary charging connections.

SUSPENSION - The large increase in vehicle weight due to the extra load of the HEV conversion required the suspension to be altered. The original springs did not provide adequate jounce. The conversion weight bias of 48% front weight/52% rear and accounting for the limit of 5% left/right bias from neutral resulted in the following added weight to ground per wheel from the original configuration: 91 kg front, and 235 kg rear. Four new springs were purchased to meet these new loads. The damping coefficient of the MacPherson strut was not modified, hence the suspension characteristics have changed to a degree. However, after discussion with a strut manufacturer, the design team feels this change should not cause any significant problems.

PERFORMANCE RESULTS

The AMPhibian has been tested on public roads and on a chassis dynamometer. Testing on public roads has included operation on highways, city streets, and rural roads. The terrain includes rolling hills in addition to level areas. The vehicle was driven on a modified FUDS cycle during testing on the dynamometer. General AMPhibian performance results are summarized in table 2.

RANGE - During zero-emissions or ZEV mode the AMPhibian operates as a traditional electric vehicle. The vehicle has a range in ZEV mode of at least 70 km, and perhaps as high as 90 km, depending upon driving conditions. Total ZEV range testing is incomplete. The AMPhibian has been tested on a chassis dynamometer in HEV mode with the APU operating continuously. Based upon data from this test, the vehicle has a projected range, in HEV mode, of 740 km. Thus, the combined range between re-charging or re-fueling of the AMPhibian is estimated to be 810 km.

Table 2. Performance results for the U. S. Naval Academy's Hybrid Electric Vehicle.

ZEV range:	70-90 km
HEV range:	740 km
Combined range:	810-830 km
Acceleration, 0 to 70 kph:	< 18 s
Gradeability:	> 6 %
ZEV Efficiency:	5.2 km/kW-h
HEV Efficiency:	1.9 km/kW-h
Total Range Efficiency:	2.3 km/kW-h

ACCELERATION - The AMPhibian can achieve an acceleration from zero to 70 kph in less than 18 seconds when operating in either ZEV or HEV mode. This acceleration rate is somewhat lower than desired due to the following reasons. First, the motor controller has ramp circuitry that does not allow large rate changes in applied motor voltage. This means that when the accelerator pedal is pressed, the motor voltage will not immediately rise, but will ramp-up. Second, the motor controller is limited to 400 amps or less depending upon controller temperature. These problems act to limit the peak torque and horsepower that can be developed in the DC motor. Finally, the stock Escort manual transmission is not ideally suited for the DC motor. The transmission exhibits a large gear ratio increase between first and second gears. This translates into a large load torque increase which the DC motor has difficulty in overcoming.

GRADEABILITY - Currently, the vehicle has been evaluated on a 6% grade. This level of grade posed no serious problem and, in fact, the AMPhibian was able to accelerate from a stand-still even when started in second gear.

EFFICIENCY - During ZEV testing the electrical energy supplied by the batteries was measured found to be 5.2 km/kW-h. While operating in HEV mode, the efficiency was found to be 1.9 km/kW-h. The measured HEV efficiency is somewhat low as the APU stalled three times during the test and had to be operated with the choke on. This problem occurred due to a mechanical problem with the APU speed governor.

A total combined efficiency can be found through a weighted average of the two operating modes where the weighting factor is the vehicle range while operating in each mode. This yields a total range combined average of 2.3

km/kW-h. Obviously, the combined efficiency will vary significantly depending upon the operating strategy used when the vehicle is operated on a daily basis since the typical distance between re-charging/re-fueling will be significantly less than the total combined range of 810 km. For example, the vehicle could be operated solely in ZEV mode if the typical distance between re-charging was less than about 70 km. This would provide an effective efficiency that is much higher than a vehicle which is operated for long distances.

EMISSIONS - The AMPhibian emissions were measured during the modified FUDS test in HEV mode with the APU on. These results are presented in table 3. To help minimize the emissions, a three-way catalytic converter is installed near the exhaust manifold of the APU. Additionally, fresh air is inducted into the exhaust just upstream of the converter to aid in the oxidizing process. Unfortunately, due primarily to the stalling problems which occurred during the emissions test as mentioned previously, the AMPhibian did not meet the TLEV standards. The carbon monoxide emissions were approximately ten times over the standard. This is due to the engine stalling problems as well as the resultant operation while under choke. The rich air-fuel mixture would naturally produce higher levels of carbon monoxide due to incomplete combustion. The mechanical problem with the speed governor has been fixed; however additional control work is required to prevent engine stalls under heavy loads when the propulsion batteries are significantly discharged. Further, a larger air-pump is being considered for fresh air induction to the converter.

SUMMARY

The design of a feasible hybrid electric vehicle for use in near-term applications has been presented. Continued testing and evaluation will reveal the reliability and durability of the various system components. However, the chosen batteries are expected to only maintain peak performance, under normal daily use, for 18 to 24 months before requiring replacement at an approximate cost of $1500. This cost is offset somewhat by the slightly reduced operating expenses due to the reduced use of gasoline, assuming electric power discounts for charging. Another costly system component is the lightweight alternator used in the APU. This item cost about $5000. The cost of this component would be greatly reduced if it were mass-produced. The total cost of components(less safety items and including a credit for the stock engine) came to about $14,000. So the cost of the alternator is a major portion of the total cost. Obviously, since the standard Escort cost about $10,000, the conversion is not yet a cost effective alternative to existing gasoline vehicles. However, the potential reduction of smog in urban areas will continue to dictate the use of these vehicles.

Future vehicle enhancements include the addition of small alternators to be used in conjunction to the existing brake system, to provide regeneration. Simulations provided by other authors(Wyczalek, 1992) have shown potential energy savings from 6 to 20%, depending upon the driving conditions. Additionally, a significant effort is underway to both determine and implement a better control strategy. Further, the mechanical speed control of the APU engine may be replaced with an electronic control to help prevent engine stalls under heavy acceleration with significantly discharged batteries.

The development of this vehicle has proven to be a valuable lesson in engineering design for both the midshipmen and the faculty of the U. S. Naval Academy. The team is currently working on design modifications and refinements in anticipation of the Second Hybrid Electric Vehicle Challenge, co-sponsored by Saturn Motors Corp., SAE International, and the Department of Energy. This event is scheduled to be held during June of 1994.

REFERENCES

Wyczalek, F. A., and Wang, T. C., SAE Tech. Paper, 1992, 920648.

Johnson Controls, Inc., Specialty Battery Division, Form number 41-6416(rev 5/92).

Table 3. U. S. Naval Academy HEV Emissions.

Component (gm/mi)	AMPhibian	TLEV Std
NMOG	0.111	0.125
CO	36.8	3.4
NOx	0.06	0.4

940338

The Selection of Lead-Acid Batteries for Use in Hybrid Electric Vehicles

Frank C. Madeka, Gregory W. Davis, and Gary L. Hodges
United States Naval Academy

ABSTRACT

Lead-acid batteries are currently the least expensive option for use in hybrid electric vehicles. The battery selection process for use in hybrid electric vehicles is complicated due to the limited use of these vehicles. Considerable data exists for the use of lead-acid batteries for other purposes. Unfortunately, much of this data is not directly applicable when these batteries are to be used in hybrid electric vehicles. Currently, there exists a wide variation in the type and format of battery data that is provided by the manufacturers. A comprehensive survey of deep cycle lead-acid batteries was conducted. Batteries were compared using manufacturers published data. To provide a consistent basis for comparison, various performance parameters-energy capacity, capacity, weight, and volume- were normalized with respect to both weight and volume. Additionally, all data was normalized to reflect three hour discharge rates. The battery selection process is then described in detail. Finally, recommendations are presented for the type and format of data that, if provided by the manufacturers, would greatly assist the hybrid electric vehicle battery selection process.

INTRODUCTION

At the turn of this century, nearly half the cars in the United States were electric cars[1]. However, as internal combustion engines improved, electric cars have become engineering curiosities. Ironically, at the turn of the next century, electric and hybrid electric cars, in particular, are again being strongly considered. Environmental legislation is driving this renewed interest.

In 1998 2% of all vehicles delivered in California by manufacturers selling 35,000 vehicles per year must be "zero emission." In 2001 and 2003, the requirement increases to 5% and 10%, respectively. Two other states have adopted the same rule and nine more states and the District of Columbia are preparing similar legislation. These "zero emission' vehicles are presumed to be electric, as power from batteries does not produce NO_x, CO or hydrocarbons. Emissions associated with the power generation necessary to charge batteries are not considered and are covered by other legislation.

BACKGROUND

Historically, a significant problem with electric cars is their limited range. This limited range is overwhelmingly due to the limited energy density of existing batteries when compared to the energy density of conventional petroleum based fuels (e.g., gasoline). To overcome this battery limitation a supplemental power source can be added to extend driving distances, making the car a hybrid electric vehicle (HEV). This is presumed to be an interim solution until batteries are developed with energy densities high enough to facilitate consumer accustomed driving ranges. Until that time, a typical, daily Los Angeles commuter, for example, could easily drive to and from work by battery power alone assuming a 40 mile range is needed. However, that same commuter could supplement the battery capacity by extending the range via a supplemental power source thereby allowing weekend trips to destinations like Las Vegas.

Data needed to evaluate batteries for hybrid electric vehicles (HEV) depends upon how close the vehicle resembles a conventional internal combustion powered, or a pure electric vehicle. For example, if the electric hybridization is one of power-assist, the battery data needed tends to reflect a starting battery since the discharge rates are typically of short duration. Power-assist hybrids have shorter zero emission vehicle (ZEV) range with the stored electric energy used to boost the conventional internal combustion power source for short intervals. However, for a range-extender hybrid, the battery data needed tends to reflect deep cycle batteries. Range-extender vehicles have significant ZEV range with an auxiliary power unit (APU) used to supplement the electric motor for short intervals. Range extenders have an inherent advantage of best meeting the intent of the environmental legislation. Since hybrids can be designed for <u>any</u> combination or degree of either range-extension or power-assist, use of existing battery specification data has limited direct applicability to HEVs. This forces HEV designers to resort to extrapolating available data, further testing or engineering judgment.

The United States Naval Academy (USNA) participated in the Ford Motor Company, SAE International, United States Department of Energy sponsored 1993 Hybrid Electric Vehicle Challenge. Converting a 1993 Ford Escort LX Wagon to a series HEV. The series approach involves successive connection of individual components. That is, the auxiliary power unit (APU) supplements the battery which, in turn, then powers the electric motor which then moves the vehicle. Whereas, the parallel approach is one where multiple paths exist to transmits mechanical power to the ground. This international, intercollegiate competition between 30 North American universities highlighted several short comings in published battery specification data for HEV designers. USNA designers desired an inexpensive, readily available, small, lightweight, easily rechargeable battery having high specific power and high specific energy with the following technical characteristics:

1) battery stack voltage of 120 volts (motor controller constraint)

2) maximum of 20 kW-hr capacity at a 3 hr discharge rate (HEV Challenge constraint)

3) total battery weight less than 500 kg (GVWR constraint)

4) minimum of 300 recharge cyclic life at 80% depth of discharge (HEV Challenge requirement)

The USNA HEV, named AMPhibian, operates with two battery systems. The first is a 12V system which is used to power the lighting and accessories. The second is a 120V battery stack which powers the prime mover (DC motor) and supplies power to recharge the 12V battery.

The USNA design team selected a 120 volt battery stack based on the MOSFET motor controller rating of 120 volts. This enabled USNA to determine an order-of-magnitude calculation of the cost of batteries having characteristics considered to be superior to those of conventional lead-acid batteries. Results of this analysis lead the design team to limit battery selection considerations to off-the-shelf lead-acid batteries. For example, Nickel-Iron batteries were found at a cost of $1800 per six volt battery or $36,000 for a 120V battery stack. Nickel-Cadmium were found at a cost of $964 per six volt battery or $19,300 for a 120V battery stack. Both estimates far exceeded AMPhibian budget constraints, hence, the self-imposed limitation to lead-acid batteries.

LEAD ACID BATTERY SELECTION

The lack of published, comprehensive, independently verified, technical battery performance data covering an extensive number of battery models and manufacturers complicates the task of battery selection. The designer is constrained to comparing only those characteristics typical available. Typical manufacturer supplied data includes the following:

- dimensions
- cranking amps (cold (CCA) and/or marine (MCA))
- weight
- reserve capacity and/or amp-hr rating
- battery categorization (e.g., starting, deep cycle)

Typical manufacturer supplied data is seen in Tables 1 and 2.

HEV designers can perform only limited comparative analyses based upon this information. This limited analysis can be divided into two groups; 1) direct evaluation and 2) relative performance evaluation.

DIRECT EVALUATION

Using the readily available data, some direct evaluations can be made to eliminate those batteries that

Table 1. Typical Manufacturer Supplied Data (Trojan Battery Company)

TYPE	NO. OF PLATES	C.C.A.[A]	M.C.A.[C]	RESERVE CAPACITY[B] MINUTES @ 25 AMPS	75 AMPS	20 HOUR RATE[D]	PLT. QTY.[E]	COVER/ CONT. COLOR	DIMENSIONS/INCHES[G] L	W	H
MARINE STARTING—12 VOLT—SEA STALLION											
24SM-2	54	420	515	65	–	–	60	BK/TW	11¼	6¾	9¾
24SM-3	66	520	635	85	–	–	60	R/TW	11¼	6¾	9¾
24SM-4	90	700	860	125	–	–	60	TW/TW	11¼	6¾	9¾
Deep Cycle[H]											
MARINE DEEP CYCLE—12 VOLT—SEA STALLION											
24TMS	66	400	490	125	–	72	60	R/TW	11¼	6¾	9¾
24TM	78	520	635	135	–	85	60	R/TW	11¼	6¾	9¾

Table 2. Typical Manufacturer Supplied Data (GNB Incorporated)

MARINE LINE

BCI GROUP SIZE	DIMENSIONS (INCHES) LENGTH	WIDTH	HEIGHT	GNB SPECIALTY TYPE	PEAK CAPACITY (MINS.) AT 80°F AT 25 AMPS	CRANKING PERFORMANCE AMPERES AT 0°	MARINE CRANKING AMPERES AT 32°	APPROX. SHIPPING WEIGHT LBS. WET CHARGED	PRO RATA SERVICE ADJUSTMENT IN MONTHS
STOWAWAY® MARINE/RV–12 VOLT–SEALED TECHNOLOGY/DEEP CYCLE/MARINE STARTING									
27	12¾	6 15/16	9 11/16	ST-800[A]	160	600	800	60	36[B]
24	11	6 15/16	9 11/16	ST-675[A]	120	500	675	48	36[B]
"ACTION PACK"™ MARINE, RECREATION & SPECIAL APPLICATION–12 VOLT–DEEP CYCLE									
27	12¾	6 13/16	9¼	AP-115[A]	160	-	N/A	55	36[C]
24	11	6 13/16	9¼	AP-80[A]	120	400	N/A	42	36[C]
U1	7⅞	5⅛	7⅞	AP-34[D]	42	200	N/A	20	36[C]

cannot meet basic design requirements or constraints. Direct comparative evaluations for a given stack voltage included the following:

capacity
total weight
total volume

Evaluation of the capacity is dependent on the battery stack arrangement. Since the battery stack voltage was constrained to the motor controller rating of 120 volts and the HEV Challenge imposed a 20 kW-hr battery constraint, the maximum battery or combination of batteries amp-hr rating could not exceed 167 amp-hr or:

$$\frac{20kW}{120V} = 167 amp-hr$$

Since USNA desired to maximize ZEV range, the maximum capacity was also desired. Therefore, capacity was found by not just simply placing the batteries in series to achieve a 120 volt stack, but also by paralleling batteries with relatively small amp-hr ratings to best achieve the maximum amp-hr derived constraint at 120 volts. In short, a given battery stack voltage can have an infinite number of incremental variations in the capacity depending on the number of batteries placed in parallel within the stack. This also means that smaller amp-hr rated batteries *should not* be immediately discounted when compared to the derived rating. While, those batteries with an amp-hr ratings higher then the desired rating for a given voltage can, to some extent, be discounted since it is not practical to reduce the amp-hr rating of a single battery (i.e., cannot divide a given battery).

With this basic understanding, USNA looked at over 50 different lead-acid batteries arranged into a 120 volt stack. The direct evaluation of the capacity using available data at the 20 hour discharge rate can be seen in Figure 1.

Because this figure is based upon available data, it provides only the 20-hour discharge rating. It does not provide the data needed to evaluate the capacity for the shorter discharge time intervals encountered by an HEV. Also, the readily available data does not cover the 3-hour discharge rate specified by the HEV Challenge sponsors. As such, one approach to get the estimated 3-hour rate based upon the 20-hour rating is to use a *rule of thumb*, known within battery circles. That is, an estimate of the 3-hour discharge rating can be made by taking 74% of the 20-hour rating. Thus Figure 1 is modified as shown in Figure 2.

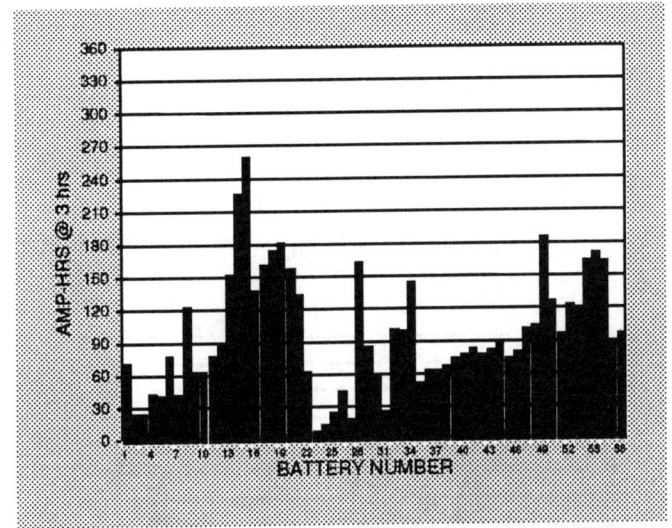

Figure 2. Est. Battery Capacity @ 3 hour Discharge Rate

Using this modified data, several batteries can be eliminated since the 20 kW-hr capacity at the 3-hour discharge rate would be exceeded. However, due to the inaccuracies, this *rule of thumb* is inadequate for HEV designers. For example, the battery eventually selected by the USNA had some additional data provided by the manufacturer. This data is plotted in Figure 3, which shows discharge rate versus time.

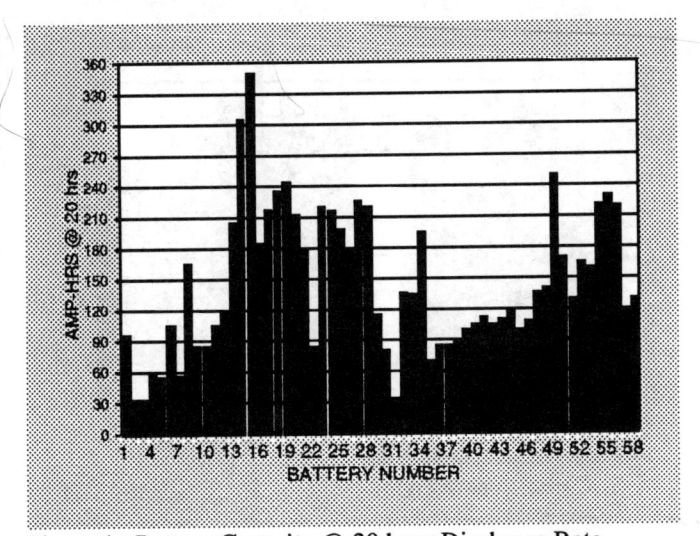

Figure 1. Battery Capacity @ 20 hour Discharge Rate

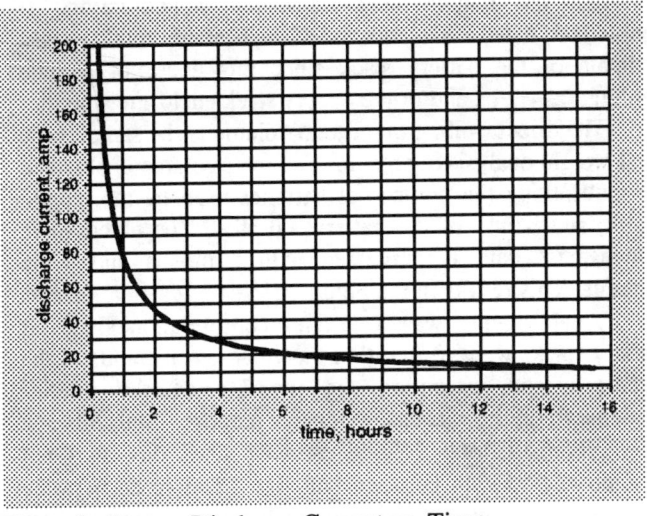

Figure 3. Battery Discharge Current vs. Time.

Based upon figure 3, the actual 3-hour rating is 105 amp-hr. This compares to 122 amp-hr using the *rule of thumb* or approximately a 15% error. The error developed by this *rule of thumb* estimate is compounded since the actual discharge rates in the range-extender HEV are less than 3-hours. In short, published battery data must be expressed such that Figure 3 can be generated in order to best match the HEV design being evaluated. Thus, some batteries may have been erroneously eliminated from consideration since the estimated 3-hour rating was higher then actual.

The other direct evaluation analyses, total weight and total volume, are easily generated from available data. The total weight is accomplished by multiplying the number of batteries arranged within the stack by their respective individual wet weight. It should be noted that the typical, available data expresses both dry (shipping purposes only) and wet weights. The result of this direct evaluation can be seen in Figure 4.

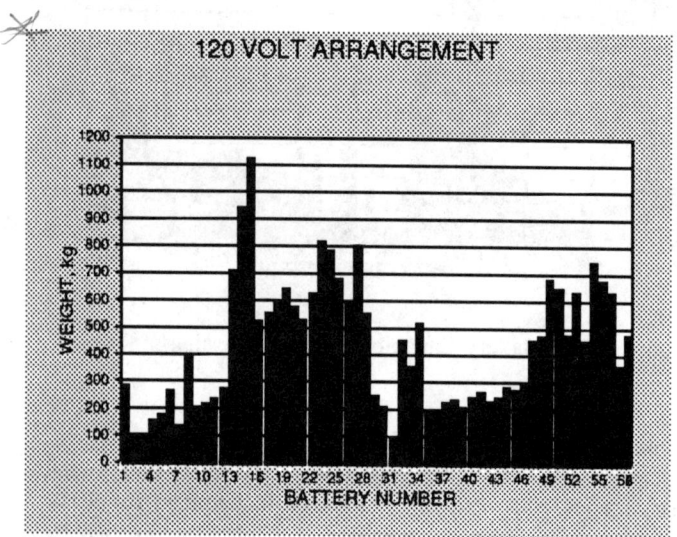

Figure 4. Total Battery Stack Weight.

Using this data numerous batteries can be eliminated since the total stack weight would exceed the 500 kg constraint based upon the gross vehicle weight rating. It should be noted that this is an optimistic total weight estimate of the battery stack since this does not include cabling, fasteners, and/or the battery stack enclosure weight.

The direct evaluation of total volume is the same as the total weight analysis approach. The total volume is obtained by multiplying the number of batteries used by the length, width, and height of the battery. It should be noted some dimensional data is expressed with terminals and some without. This results in the Figure 5.

Total volume was a tertiary consideration since all battery stacks were well within the volume constraints of the Escort wagon, assuming the majority of the interior passenger volume was usable for this purpose. Still, this is a *very optimistic* total volume estimate since this does not assume any spacing between batteries. The space between batteries is not only needed to facilitate air flow to purge hydrogen gas generated during charging but more importantly, and much overlooked in HEV design, is the

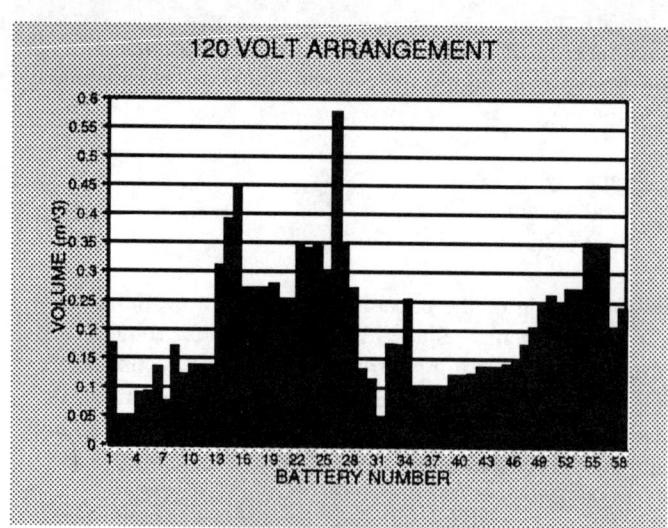

Figure 5. Total Battery Stack Volume.

space needed to remove the significant amount of heat generated during charging and discharging. Had the interior cargo space not been utilized in the Escort wagon, total volume, including cooling passages would have been a substantial design constraint.

RELATIVE PERFORMANCE EVALUATION

The direct evaluation data can be normalized. This provides insight as to the relative performance of those batteries remaining after the direct evaluation. The evaluations that can be performed using available data including the following:

capacity-to-weight
capacity-to-volume
volume-to-weight

This can be seen in the Figures 6 - 8.

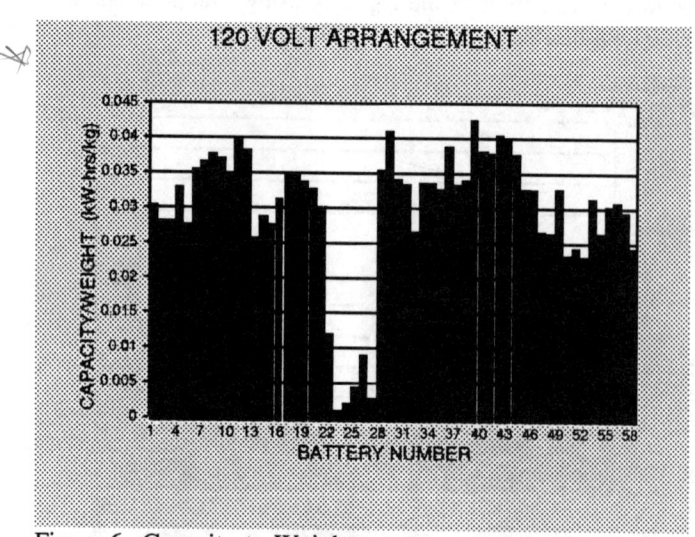

Figure 6. Capacity-to-Weight.

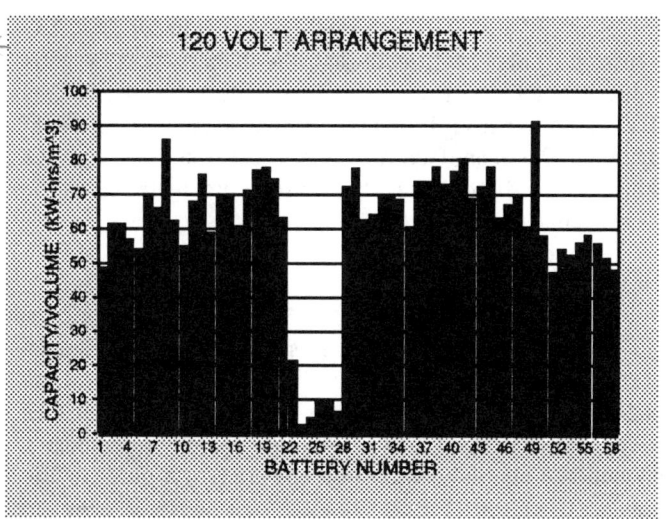

Figure 7. Capacity-to-Volume.

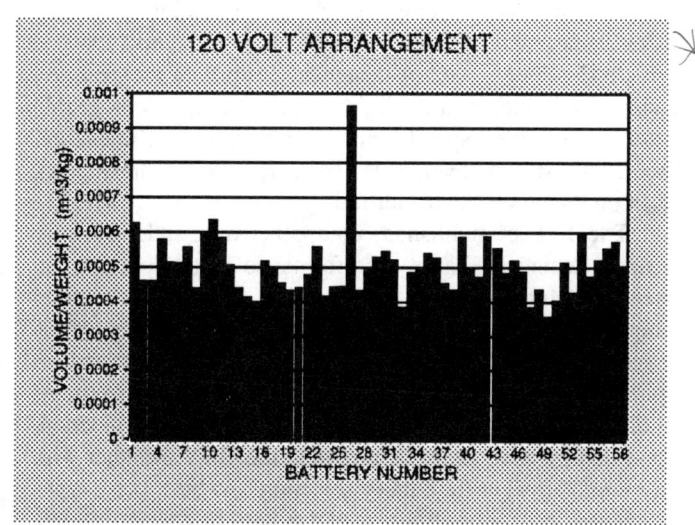

Figure 8. Volume-to-Weight.

Since these ratios provide insight into the relative performance of a given stack, the designer may place higher consideration for one given ratio than for another. For example, had volume been constrained to the outside of the passenger compartment those batteries with favorable capacity-to-volume ratios would have been desired. However, USNA was weight driven and considered those ratios more important. As expected, those battery stacks that were arranged by paralleling small capacity batteries displayed "poor" ratios since these stacks contain multiple housings. Another short coming of this evaluation is that it indicates a trend toward those stacks which are composed of batteries of higher voltages or more cells (e.g., 8v, 12v, 24v). However, this may not be the best design since it may require the removal of a multi-celled battery with only one cell "bad." In short, it is desirous to have a smaller number of cells per battery to facilitate removal and replacement of the least number cells -- ideally just one cell. Finally, this evaluation is limited due to the optimistic nature of the direct evaluation since cabling, fasteners, stack enclosures, and allowance for ventilation passages are not accounted for.

OTHER EVALUATION DATA NEEDED

The preceeding comparative analysis provides the HEV designer with only limited insight for battery selection. At best, this type of analysis can only eliminate those batteries which clearly do not meet the design requirements or constraints. For example, numerous batteries were eliminated because the total weight was greater than 500 kg and/or would exceed the 20 kW-hr maximum capacity. However, the analysis does provide some normalized information thus focusing the designer to those batteries with favorable capacity-to-weight, capacity-to-volume, and volume-to-weight ratios.

BATTERY LIFE - Evaluation of those batteries meeting or exceeding the minimum 300 recharge cyclic life at 80% depth of discharge (DoD) is limited to qualitative assessment only. The typical manufacturer published data lacks sufficient information to determine the estimated cyclic life of the battery. The qualitative assessment was based upon the *gross assumption* that batteries categorized as "deep cycle" would meet the requirement. However, for some batteries, a limited amount of life data is available. For example, Table 3 shows battery cycle life at three levels of depth of battery discharge.

Table 3. Brochure Life Cycle Information (Johson Controls)

		Nominal Voltage	
		12 volt unit	5 volt unit
Constant current discharge at various rates:	8 hour	9.1	18.2
	24 hour	3.5	7.0
	100 hour	.9	1.9
Cycle life at:			
80% Depth of Discharge		900	900
50% Depth of Discharge		1800	1800
40% Depth of Discharge		2150	2150
Internal Resistance (fully charged)		5.0 milliohms	2.0 milliohms
Operating Temperature:	Charge	-4 to +140 degF	-4 to +140 degF
Weight		69 lbs.	69 lbs.

This can be subsequently be extrapolated and plotted, Figure 9. With this data, HEV designers can estimate battery life.

SPECIFIC POWER - The USNA design team was unable to evaluate the specific power of a given battery for either discharging or charging. The USNA team was not able to obtain or extrapolate any data to quantify the ability of a given battery to charge or discharge. The significance of this data cannot be overstated. Qualitatively a battery selected may have a significant amount of stored energy yet the ability to discharge and charge rapidly are equally important to HEV designs. For example, regenerative braking, which converts the kinetic energy to electric energy, can rapidly generate significant amounts of electricity. Yet it is qualitatively known that a battery is unable to fully store this magnitude of energy in such a short period of time. Similarly, it is desirable to achieve high energy charging rates to minimize charging times. Yet there is no quantitative data to evaluate which batteries can effectively and efficiently accomplish this. Discharging data is likewise needed. For example, during acceleration a large current

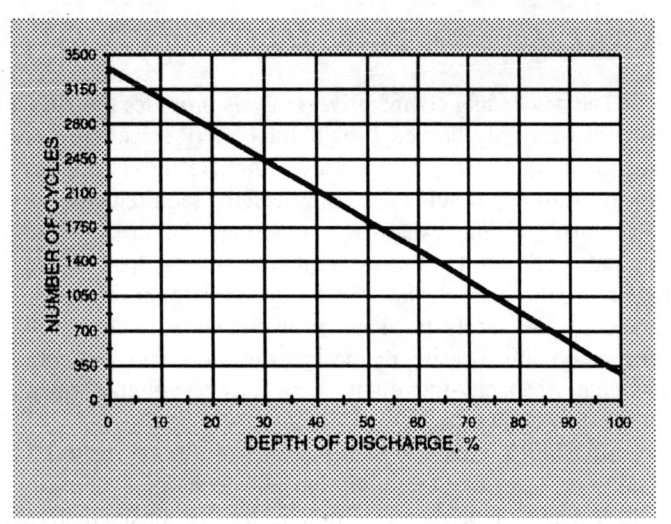

Figure 9. Estimated Cycle Life versus Depth of Discharge.

draw is needed. Although the average power can, to some extent, be determined from the Figure 3 which shows discharge rate versus discharge time this is not adequate for the short charge/discharge times. Thus instantaneous power is needed to truly evaluate a given battery. An alternative maybe a quantified value(s) for the batteries internal impedance, both charging and discharging. Since this data is not readily available, this area of concern was not addressed by the design team.

TEMPERATURE - It is known that battery performance, capacity and life are dependent on the temperature. Typically, higher temperatures yield higher capacities and shorter battery lives. Since HEVs have a large number batteries within a given battery stack, the variation of temperatures can vary widely within the stack. As such, the batteries within the stack are being utilized to varying degrees. Coupling this information with the stack configured in a series configuration, the stack then is only as good as its' weakest link. Yet the design team was not able to obtain temperature characteristics for a specific battery for comparative evaluations. Instead, the design team had to resort to generalized lead-acid data and assumptions. That is, all lead-acid batteries were assumed to behave the same. Intuitively, this does not appear to be a prudent assumption.

STATISTICAL CONFIDENCE - The USNA design team could not statistically weight average a given battery characteristic from a given manufacturer for comparison to another manufacturer. That is, none of the readily available information provided any statistical information. It is not known if the data reflects, for example, nominal values, $+3\sigma$, -3σ, or other statistical condition. Therefore, the design team had to assume all data was nominal and equal in all other statistical respects. Intuitively, this too does not appear to be a prudent assumption.

FINAL SELECTION

Based upon the above direct and indirect evaluations and discussions with EV enthusiasts,[2] and professional EV converters[3], the USNA selected battery number 8. The discussions only confirmed that the battery was a good choice but not necessarily the best battery for the USNA HEV. Additionally, the USNA design team was forced to rely upon the manufacture's verbal confirmation that the battery selected would meet the minimum recharge cyclic requirement.

SUMMARY

Environmental legislation is spurring renewed interest in electric and hybrid electric, in particular, cars. Data needed to evaluate batteries for hybrid electric vehicles (HEV) depends upon how close the vehicle resembles a conventional internal combustion powered, or a pure electric vehicle.

The United States Naval Academy (USNA) participated in the Ford Motor Company, SAE International, United States Department of Energy sponsored 1993 Hybrid Electric Vehicle Challenge. USNA designers desired an inexpensive, readily available, small, lightweight, easily rechargeable battery having high specific power and high specific energy. The USNA design team selected a 120 volt battery stack based on the motor controller rating.

The lack of published, comprehensive, independently verified, technical battery performance data covering an extensive number of battery models and manufacturers complicates the task of battery selection. The designer is constrained to comparing only those limited characteristics typical available. Analyses can be divided into two groups; 1) direct evaluation and 2) relative performance evaluation. These comparative analyses provided the HEV designer with only limited insight for battery selection. At best, these types of analyses can only eliminate those batteries which clearly do not meet the design requirements or constraints.

Further data and evaluations are needed to adequately select batteries. For example, the typical manufacturer published data lacks sufficient information to determine the cyclic life of a battery. Also, the USNA design team was unable to evaluate the specific, instaneous power of a given battery for either discharging or charging. Temperature dependent characteristics for a specific battery are not available for comparative evaluations. Finally, the USNA design team could not statistically weight average battery characteristic from manufacturer to manufacturer. That is, none of the readily available information provided any statistical information.

REFERENCES

1 Noel Perrin, "Solo. Life with an Electric Car," W. W. Norton & Company, New York, NY, 1992

2 Mr. David Goldstein, President, Electric Vehicle Association of Greater Washington, D.C., private conversations.

3 Mr. Douglas Cobb, President, Solar Car Corporation, private conversations.

940339

Development of the University of Alberta Entry in the 1993 HEV Challenge

M. D. Checkel, V. E. Duckworth,
C. B. Collie, and K. M. W. Workun
University of Alberta

ABSTRACT

Because of the limitations of their storage batteries, electric cars have always suffered from short range, high weight, and high cost. New battery technologies will provide a significant improvement but all-electric vehicles will still tend to be heavy, costly, and severely limited in range compared with their combustion-engined counterparts. Despite these inherent disadvantages, there is a huge impetus for electric car development because of the pollution disadvantages of the combustion engine. Given the weight/cost/range problems of purely electric cars, it is desirable to develop hybrid cars which have the capability of operating as zero-emission electric cars in urban areas and which use a small internal combustion engine to extend the operating range. The internal combustion engine and its fuel are far lighter, cheaper, and more effective at extending range than carrying enough battery capacity to give an all-electric vehicle a suitable range.

The U.S. Department of Energy, Ford, and SAE organized a student design competition to highlight the possibilities of hybrid electric cars. The University of Alberta, along with 29 other North American university teams, spent eighteen months developing and building safe, practical, road-licensed cars with hybrid electric drive systems. The car developed by the University of Alberta team demonstrates the near-term feasibility of the hybrid electric concept and was successful in winning the 1993 HEV Challenge Competition. This paper describes the major design choices and the development process used to produce the University of Alberta vehicle.

INTRODUCTION

The challenges of developing a fully functional, practical electric automobile have been around since the first electric vehicles were challenging the early gasoline and steam vehicles for supremacy. At that time, internal combustion engine capabilities developed rapidly while there was a lack of progress in developing suitable storage batteries for the electric cars. The result has been a century of domination by combustion-driven vehicles. Now, however we are finding the limitations on using internal combustion engines to power virtually the entire transportation system in large cities. The problems became obvious in sensitive areas such as the California coastline more than four decades ago and led to controls on vehicle emissions. Rigorous pollution controls now limit new vehicles to less than 10% of pre-control emission levels. However, with growing population, increasing size of urban areas, increasing use of personal transportation, and increasing traffic congestion, the problem simply cannot be eliminated. For example, despite the continuing renewal of the California vehicle fleet with ultra-clean vehicles, transportation continues to dominate the pollutant emissions in California. In fact, despite the tighter emission standards, the total emissions in California, which have been declining steadily since the 1970's are expected to bottom out and start increasing again due to continuing population and urban traffic increases into the next century [1]. Even urban areas outside California, which were saved from serious vehicle pollution problems by a climate which effectively diluted the pollution, have now grown to the point where the pollution can no longer effectively dispersed and are seeing increased smog, haze and ozone.

This results in a renewed interest in electric cars to provide clean transportation in the congested cities of the twenty first century. Electric vehicles can, in theory, use relatively clean nuclear or hydro power for transportation and thus virtually eliminate the combustion-generated pollutants. Even if the electric power is generated by oil or coal combustion, the pollutants can be generated outside the city centre, minimizing the direct population exposure and reducing the contribution to the urban smog problem. Also, pollutants generated at a point location like a power plant can have sophisticated, capital intensive pollution controls applied with a reasonable level of success.

Are electric vehicles ready? Enormous development progress has been made in electric motors, motor controllers, and storage batteries over the past century. However, the energy storage density of an electric battery is still pathetic compared with the energy storage of a similar mass or volume of hydrocarbon fuel. Table 1 shows some practical energy

Table 1.
Typical On-Board Energy Density of Fuels and Batteries (Incl. Tankage, Mounting Hardware, etc.)

FUEL OR BATTERY	ENERGY DENSITY kJ/L	kJ/kg
GASOLINE	33,000	40,000
METHANOL	16,000	18,000
NAT. GAS	8,000	6,000
Na-S	72	355
Ni-Cad	36	111
Pb-Acid	18	72

Batteries: Na-S = Sodium Sulphur, Ni-Cad = Nickel Cadmium, Pb-Acid = Lead Acid

storage densities for vehicle applications. The fuels have a huge advantage over the batteries in total energy storage. This is not surprising considering that the fuel energy is released by reacting that fuel with several times its mass of air, utilizing all the chemical bond energy available in the combined mass, and then dumping the fuel and its products into the atmosphere. By contrast, battery energy release involves chemical reaction of only a small fraction of the battery mass, with no outside reactants, and the chemical reactions are ones which can be reversed simply by applying a reverse voltage across the reactants when they are returned to a charging station! It can be argued that the electrical energy stored in a battery is converted to propulsive power much more efficiently than the chemical energy in a fuel. However, this still only reduces the effective propulsive energy of a fuel by four or five times in comparison with a battery which contains two or three orders of magnitude less energy.

To illustrate the problem, consider that a typical 1100 kg car requires about 35 L of gasoline for a 500 km driving range. The fuel and tank adds a negligible mass and volume to the existing vehicle. Only about 20% of the energy in the fuel is used and the rest is dumped as cooling system and exhaust heat. By contrast, a highly efficient vehicle which converts the battery energy to propulsive energy with 90% efficiency would need about 725 kg of sodium-sulphur (Na-S) battery or 2300 kg of nickel-cadmium (Ni-Cad) battery to contain enough energy for the same driving range. And, since the battery is so large and heavy compared with the car, the whole vehicle would have to be made larger, requiring more energy and hence more battery power to travel the same distance. The result is that electric vehicles are normally highly specialized to accommodate their power source. They use light weight materials to allow for battery weight and improve energy efficiency; they have limited interior space due to battery volume; and they still have a very short range of continuous operation compared with a fuelled vehicle.

Another significant difference between battery energy and fuel energy for a vehicle is the rate at which it can be replenished. Typical battery recharging schedules call for six to ten hours of charging at maximum energy input rates of a few kW. Filling up with gasoline replenishes a vehicle's energy supply at a rate measured in tens of megawatts! For a commuter vehicle with only limited daily use, this difference can be ignored since there is adequate time available for overnight recharging. However, most commuter vehicles are periodically required to do more than just routine commuting.. Examples would include running a series of errands after work, driving up to the mountains for a weekend, or driving across the country for a holiday. A vehicle which requires several charging hours per hour of operation is fine for commuting but it simply can't fulfil these occasional requirements. That makes it much less useful to a typical commuter and limits the possible market penetration of purely electric cars severely. Some future developments to resolve this problem are foreshadowed by current research. Solutions may include quick recharge batteries and high power charging stations or removable battery packs and battery exchange stations. However, the commercialization of those systems and the enormous infrastructure development which would be required to make them practical on a large scale is still far in the future.

A more feasible solution for near term introduction of electric automobiles is to develop hybrid electric vehicles. Such vehicles could operate as purely electric vehicles over a range adequate to cover most normal daily operation. Then, rather than carry enough extra batteries to provide normal vehicle range, the vehicles would carry a small internal combustion engine adequate to run the vehicle. This allows normal operation without the worry of stranding the vehicle when the battery runs down. It also allows virtually limitless extended range operation similar to normal automobiles because the vehicle can fuel up in the normal way during days of continuous operation or on cross-country trips. What is a reasonable range for electric power? Studies of commuter driving patterns show the average private vehicle will drive less than 65 km (40 mi.) in one day approximately 77% of the time. A pie chart with a typical daily driving range breakdown is shown in Figure 1.

The U.S Department of Energy, SAE, and Ford Motor Company have created the Hybrid Electric Vehicle (HEV) Challenge. This is a student design competition to develop automobiles that are capable of running a "reasonable" distance on battery power and extended distances using conventional liquid fuels, (gasoline, M85 or E100). This report details the technical design of the University of Alberta's hybrid electric vehicle, "EMISSION IMPOSSIBLE".

The competition allowed Universities to either build a complete vehicle from the ground up or to convert an existing small station wagon, the 1992 Ford Escort. The U of A team chose to convert the Escort station wagon to hybrid drive. Throughout the design process, the emphasis of the project team was to develop a car which would demonstrate the near-term practicality of hybrid electric vehicles. That meant the

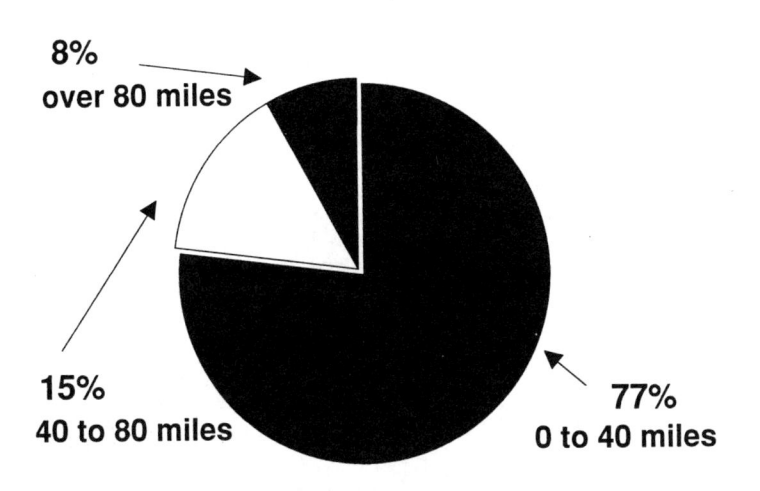

Figure 1 Typically, a private vehicle's daily driving range is less than 65 km (40 miles).

[Pie chart: 77% 0 to 40 miles; 15% 40 to 80 miles; 8% over 80 miles]

car, while meeting the rules and intents of the competition, should also be attractive to every segment of the automotive business. It should be easily and economically manufactured, should be easy to sell (to a significant target market who are currently driving gasoline-powered vehicles), and should also be easy to service. The paradigm was that this "car of the future" should look and feel like it belonged in a local driveway in 1993. This was one reason for choosing to convert the Escort wagon rather than building a futuristic automobile. Another consequence of this philosophy is that most of the components in the car are readily available production items and the car has retained all of the interior trim and other appointments consistent with a marketable vehicle. Another illustration of the philosophy is that the U of A car uses a 240 Volt, 20 Amp supply for battery charging and regular unleaded gasoline for its combustion engine. The option of using a higher electrical voltage for greater efficiency was rejected since most homes and businesses are only supplied with 240V power. The option of using alcohol fuels for emission reduction was likewise rejected on the basis that the fuels are not commonly available. The objective was always to try and match the capability, flexibility, and ease of use of the base car while adding a Zero Emission Vehicle (ZEV) operating capability.

The first section of this paper covers the power train design strategy, component selection, and integration. This is followed by a section on operating modes, drive train control strategy, emissions control, and fuel and electrical consumption of the hybrid power plant. The next section of the report covers the design of power train mounts, battery box, suspension modifications, and other necessary changes to the car. The last section of the report summarizes the features of the car and its achievements.

POWER TRAIN CONFIGURATION

There are two primary coupling methods available when configuring the power train of a Hybrid Electric Vehicle: series and parallel.

SERIES CONFIGURATION - This configuration is common to many locomotives, mine trucks, and commuter buses. The premise is simple. An internal combustion engine drives an electric generator which supplies the required electrical power to an electric drive motor which actually drives the vehicle. A battery system may be used to store excess electrical output.

The main advantage of this system is its simplicity. The power transfer between the combustion engine and the electric drive motor is independent of each respective unit. The internal combustion engine can be sized and tuned to operate at its optimum speed and efficiency while supplying the average required vehicle power. The battery system can supply excess power during vehicle acceleration and absorb the available excess power when vehicle loads are low. This allows a substantial reduction in combustion engine size compared with a normal vehicle. However, because the vehicle is driven only by the electric motors, the electrical system must be oversized to give appropriate peak performance. Also, using an internal combustion engine to drive a generator, the generator to charge batteries, and the batteries to run motors is less efficient than using the internal combustion engine to supply power directly to the drive train.

PARALLEL CONFIGURATION - This configuration allows for either the electric motor or the internal combustion engine to drive the vehicle independently. In addition, the capability to drive with both combustion engine and electric motors simultaneously allows the designer to size each power system for high efficiency in normal operation while still providing high performance in combined mode driving. Driving the vehicle directly with the internal combustion engine increases efficiency in hybrid mode since there are no electrical losses in series with the engine's mechanical output. However, the parallel configuration is not as simple to implement as series since some variable mechanical linkage is required between the combustion engine, the electric motor(s), and the vehicle's final drive system.

CONFIGURATION DECISION - The series configuration appeared easier to implement. Its advantages included simplicity of component selection, simplicity of construction and installation, and simplicity of control. Its main disadvantages were the requirement for very high power electric motors for competitive performance and efficiency loss during hybrid operation. The parallel configuration required more specialized component manufacture and controls to integrate the two power systems. However, it gave higher efficiency in normal operation and higher peak performance.

Two additional factors were instrumental in making the basic power train decision. These were the predicted reliability of the power train and the acceptability of the power train operating characteristics as perceived by the driver. It was felt that reliability of the series configuration would be poorer

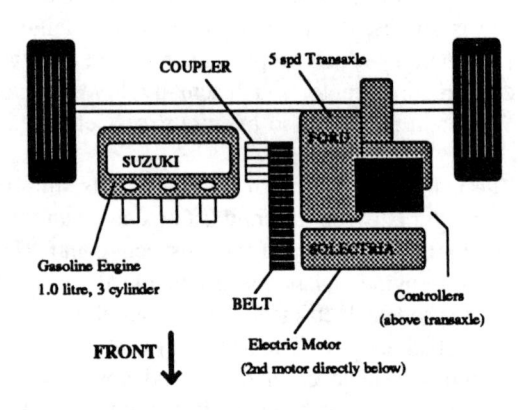

Figure 2 Schematic of Hybrid Power Train Layout for the U of A HEV. Electric motors drive clutch input via a toothed belt and the specially designed coupler

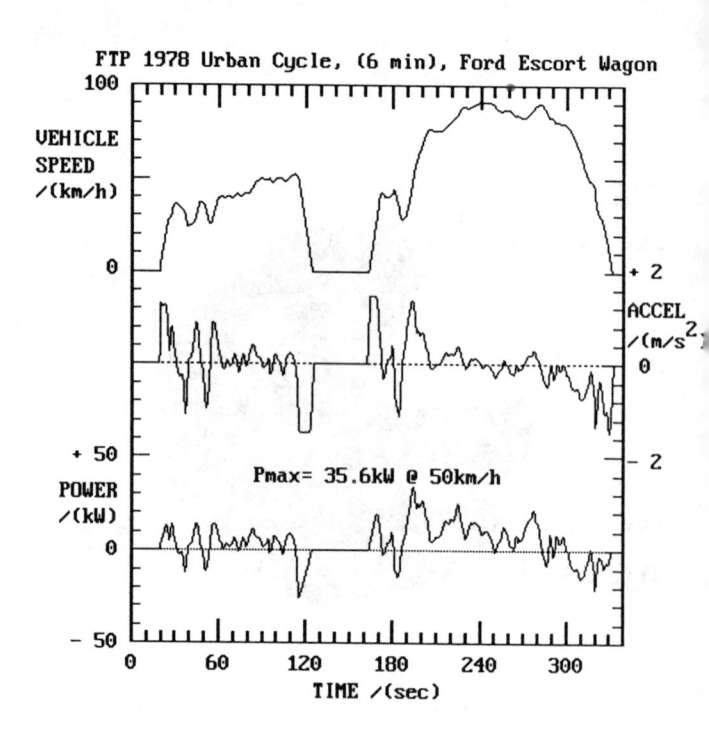

Figure 3 Part of Vehicle Simulation for UoA HEV Escort Driving FTP-78 Urban Cycle

since, by definition, every component must be functional to keep the system operating and, if any component is degraded, the vehicle can only operate at the level determined by its weakest link. In contrast, either the electrical system or the internal combustion engine could fail in a parallel hybrid system and the vehicle would remain functional on the other system. The only critical components in the parallel system are the coupler linking the engine and electric motor to the transmission and the transmission/final drive itself.

The other question was the operational characteristics of the vehicle with each drive train configuration. A car driven by a series electrical system would drive like an electric car and would have essentially the same operating characteristics whether it was in electric-only or hybrid mode. A car with a parallel system could potentially have different driving characteristics and different operating controls depending on which of three possible modes it was in. It was felt that this would be very undesirable to a typical customer but could be overcome if an effort was made to keep the driving controls and vehicle operation as consistent as possible between operating modes.

Based on the above considerations, the University of Alberta team selected a parallel hybrid electric drive for their vehicle because of higher efficiency, higher performance, and better reliability. The actual implementation of this drive, shown schematically in Figure 2, was to combine the power from two electric motors and the internal combustion engine with a coupler at the clutch input. This meant that, regardless of the operating mode, the driving power would pass through the clutch and transmission in the same fashion. This allowed the driver to perform all control operations the same as in a normal gasoline-powered car regardless of whether the vehicle was in electric-only, gasoline-only, or hybrid modes. It also allowed for the greatest possible operating flexibility.

The next question was how much power would be required. A vehicle simulation model was used to answer this question, as well as related questions about battery capacity, peak current, component duty cycle, et cetera. The model was based on a simulation program which ran vehicles through various transient cycles and calculated energy consumption, peak power demand, and other parameters of interest [2]. Basically, the model calculates the tractive power requirement due to rolling resistance, aerodynamic resistance, grade, and acceleration for each point in a second-by-second description of the operating cycle. A similar model was described by Sovran and Bohn [3]. The FTP-78 Urban and Highway test schedules [4] were used to define the minimum required vehicle performance. Figure 3 shows part of a cycle simulation with speed, acceleration, and power traces for the first two drive periods of the FTP-78 Urban Driving Cycle which lead to a minimum tractive power requirement of 36 kW. It should be noted that, on the FTP cycles, acceleration and deceleration are limited to 1.5 m/s/s. Acceleration rates during normal urban driving vary considerably but typical values rise to more than 2.5 m/s/s at low speeds [5]. The FTP cycles thus represent a minimum acceptable performance but typical driving demands more power. More aggressive driving cycle simulations including lane change accelerations, and hill climb simulations were also performed to define the desirable peak power and the required high power duty cycles. To determine the required transmission gear ratios and closely define the engine operating conditions, further simulation subroutines which included

Table 2
Acceleration Performance Comparisons of
U of A HEV and Production Escort Automobiles

	U of A 1.9L Escort	U of A EM Mode	U of A HEV Mode	Ford Escort GT
0-100m	8.2s	8.9s	7.8s	7.6s
0-30mph	3.4s	5.7s	3.6s	3.8s
0-60mph	11.5s	21.9s	12.4s	10.8s

Note: U of A vehicles tested at full GVW (3466 lb), Escort GT data based on testing from automotive magazine at curb weight.

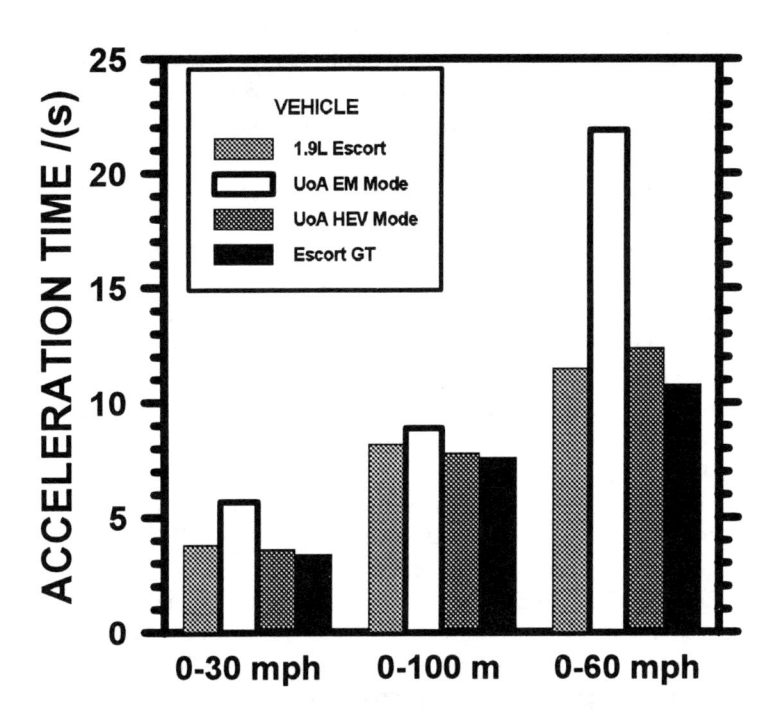

Figure 4. Acceleration Performance Comparison

engine torque limits and gear shifting were used. It became apparent that the vehicle could perform adequately with about 45 kW of tractive power but a multi-ratio transmission was highly desirable to keep the engine/motors near that peak power point over a range of vehicle speeds. It also became apparent very early that the battery requirements for a 72 km (40 mi) range would raise the Escort's operating weight to very close to the registered gross vehicle weight (GVW). For this reason, most simulations were performed assuming the vehicle was at GVW and a careful weight "budget" was used in vehicle modifications to keep from exceeding that limit.

At the competition, it became clear that the decision to use a parallel configuration and the way in which this decision was executed contributed to the University of Alberta vehicle winning several HEV Challenge performance categories and placing well in judged categories. Some of those achievements were:
  First Place in the Acceleration Event
  First Place in the Commuter Challenge Event
  First Place in the Range Event
  Second Place in the Engineering Design Event
  Fourth Place in the Efficiency Event
  Best Performance Vehicle Award
  Design Innovation & Creativity Award.
Some idea of the vehicle performance can be obtained from Table 2 and Figure 4.

ELECTRIC MOTOR SELECTION - Many different types of electric motors were investigated prior to the final selection. These included DC Shunt Wound, DC Series Wound, and AC Synchronous motors. The critical factors considered were output power, efficiency, durability, availability, weight, and cost. Some consideration was given to finding a motor or motors which could directly drive the vehicle axles. This would have given greater electric drive efficiency at some cost in operating complexity since control operations would have been different for the engine and the electric motors. In fact, no acceptable motors were identified which would give adequate torque and speed range to drive a car of this size without a transmission.

The electric motors chosen were two BRLS16, DC Brushless, Permanent Magnet motors manufactured by Solectria Corporation. These motors were relatively compact and light weight at 30 kg. Each was rated at 16 hp continuous duty with a peak power output of 22 kW (30 Hp) on a 33% duty cycle. Since continuous power is not required, this meant the vehicle would have 44 kW electric power available. Also, the motors have a peak efficiency of 94% which is critical to obtaining the optimum range with a given size battery pack.

Each of the motors has its own controller. The motors operate on three phases of modulated DC current and have a maximum draw of 220 Amps. For efficiency, the power is regulated by pulse width modulation, essentially switching the nominal battery voltage of 144V on for longer or shorter times in each of three motor inputs. The motors/controllers have an integral regenerative braking capability. Because the motors are permanently linked to the clutch input, the regenerative braking can be used in all three operating modes. Lightly operating the brake pedal switches the controllers into regenerative mode and the kinetic energy of the vehicle is converted to electrical energy by running the electric motors as DC generators. The generated power is used to recharge the battery pack. Heavier application of the brake pedal maximizes the regeneration rate and activates the normal hydraulic brakes as well.

These motors and their controllers were expensive due to both the low production volume (<200 units/year) and to the high price of developing compact, high flux density, permanent magnets.

INTERNAL COMBUSTION ENGINE - A smaller internal combustion engine was desirable for the hybrid power train for two reasons. First, a smaller engine would run more efficiently and adequate top end performance could still be obtained using the additional power of the electric motors. Another reason for a smaller engine was to provide room for the electric motors and controllers within the engine bay and to keep the total power train weight down. Various motorcycle engines were considered because of their compact layout and high power/weight ratio. However, the electronic engine management systems for the motorcycle engines would require excessive engineering effort to achieve acceptable emissions and fuel economy driving a heavy automobile. For this reason, the search was directed to small automobile engines.

After an initial search, two alternatives were investigated; the Honda VTEC engine out of the Honda Civic VX, and the Suzuki three cylinder engine out of a Swift GA (or Geo Metro). The VTEC engine was very efficient and clean burning due to its use of lean burn and variable valve timing technology. It also had a power output similar to the original Escort engine. A VTEC was procured but proved to be too large and to heavy to fit within the very restrictive space and weight budgets for the hybrid drive train.

The internal combustion engine used in the U of A HEV is the Suzuki 1.0 litre, three cylinder, four stroke gasoline engine. Because of its small physical dimensions, it was ideal for the chosen configuration, allowing both drives to be mounted within the engine bay. Additionally, it had an aluminum block and cylinder head which resulted in a weight savings of 68 kg (150 lb) as compared to the stock FORD 1.9 litre engine. The Suzuki engine had an output power rating of 41 kW (55 Hp) making its power/weight ratio 0.65 kW/kg which is significantly higher than the original engine's 0.49 kW/kg. The high specific output and inherently unbalanced three cylinder design imply reduced durability but this seems acceptable for the part-time power source of a hybrid vehicle. The Suzuki engine has a fully mapped electronic fuel and spark control system which allows it to meet California emissions standards in the small Suzuki Swift. Given the sophisticated engine control system, it was felt that the engine could be recalibrated to give adequate emissions performance with the heavier HEV.

COUPLING MECHANISM & TRANSMISSION -
The primary difficulty with the parallel configuration was the design of a coupling mechanism which would allow switching between the three driving modes; electric, gasoline, and hybrid without introducing complicated operating procedures for the driver. A coupling mechanism that met these requirements was invented, engineered and developed by the U of A student team. This coupler fits on the end of the engine crankshaft between the flywheel and the clutch. It incorporates a toothed belt drive which permanently connects the electric motors to the clutch input and a one-way clutch mechanism that connects the internal combustion engine while it is driving but slips to allow the engine to stop for electric-only mode. Because the electric motors are permanently connected, regenerative braking is possible in all driving modes.

The U of A coupler design uses a sprag type overrunning clutch mechanism normally found in the hub of an automatic transmission torque converter (Turbo Hydramatic 350). The sprag clutch, (manufactured by A1 Transmissions in Canoga Park, CA), was created to replace traditional ramp and roller clutches and is therefore of similar dimensions. The clutch consists of an inner ring and an outer ring with cylindrical elements between. When the inner ring rotates in one direction, torque is transmitted by the ramps on the inner ring wedging the cylindrical elements out against the outer ring. When the outer ring rotates in the other direction the ramps release the cylindrical elements and thus no torque is transmitted.

To optimize the electric motors for city performance and the gasoline engine for highway operation, the gear ratio incorporated in the toothed belt drive ran the electric motors at 1.4 times engine speed. This had the effect of derating the engine since its peak speed was limited to 4650 rpm rather than 6500 rpm. However, driving operations were not affected since drivers do not typically use that part of the power band and rpm range above the torque peak.

The transmission chosen for this design was the stock Escort five speed manual transmission. The five ratios were suitable for both the individual and combined drivers and the high efficiency and low weight of the transmission (less than 80 lbs) were very attractive. The computer simulations of the vehicle's zero emissions mode showed an improvement of 22.7% in the 0-100 metre acceleration time and also an 82.8 km/hr (51.5 mph) higher top speed when using the five speed transmission. The Ford transmission met all the needs of the design and was easily integrated using existing mountings, shift linkages, and axles. The transmission is located on the driver side of the engine bay (its original location) directly below the electric motor controllers.

Battery Pack - A great deal of time and effort went into the final choice of a battery. The battery pack was the single largest weight addition to the vehicle and as such, the overriding criteria in this area were weight and energy density. Additional concerns included cost, safety, physical cell dimensions, and recycling. Because the battery box occupies a large volume in the car, the box is designed around whatever cell is used and it is not a simple matter to change battery types.

An initial search showed several advanced battery types might be available. Exotic cells, such as Zinc-Air, Zinc-Bromine, Sodium-Sulphur, Nickel-Metal-Hydride, and Lithium-Permanganate were investigated to determine their usability. However, it soon became clear that they were being evaluated in prototype form. Some might not be reliable and/or would not even be reliably available. Both the Zinc Bromine and Sodium Sulfur cells were pursued vigorously. However, with the safety concerns associated with the ZnBr cell, the manufacturer was reevaluating the production of this system and would not release the technology. The NaS battery system was unavailable to us due to a contract agreement between Ford and the manufacturer and Ford's request that these cells not be given for use in the HEV Challenge. Two common battery types evaluated were Lead Acid and Nickel Cadmium. Both of these cells are significantly less energy dense than most of the advanced types currently being researched but are available in cells designed for electric propulsion.

The chosen battery system for the U of A HEV consists of STM 1.60 Nickel Cadmium cells manufactured by SAFT NIFE Corporation in France. The cell dimensions are 85 mm wide, 45 mm deep, and 278 mm tall and are specifically designed for automotive uses. It utilizes a light plastic casing and features a very high depth of discharge. This is important in an automobile where a fairly constant power availability is desirable throughout the discharge cycle. The entire battery pack is composed of 135 cells yielding a pack voltage of 170 Volts. The battery pack rating is 61 Amp-Hrs which yields a range of approximately 72 km (45 mi.) of city driving on a single charge. The total weight of the battery pack is 272 kg (600 lb). The battery pack is situated in the space originally occupied by the rear seats and will be described later in the section on vehicle structural modifications.

The cells for the battery pack were purchased at a cost over $25,000.00. As with the electric motors, this high cost is due mainly to the low production volume (<100 packs/year) and the high price of cadmium. Some additional features of these Nickel Cadmium cells which were designed for automotive use include: self-watering ability, high energy density, and recyclable internals. The self-watering system attaches to the top of each cell with reservoirs located along the sides of the battery box. These cells are approximately 1.5 times as energy dense as conventional Lead Acid cells providing 50% greater vehicle range with similarly sized battery packs. Additionally, these cells perform better than normal Lead Acid cells at low temperatures (a concern in the Canadian climate). The nickel and cadmium found inside the cells are recyclable, they can be reconstituted and reused in another cell. Only minimal gases are emitted during a normal charging cycle and NiCads are much more robust durable than Lead Acid, with less damage to the cells if they are overcharged. These cells have a life of between 2000 and 3000 charge-discharge cycles which translates into five to eight years of daily charging.

CONTROL STRATEGY

The U of A HEV design allows the vehicle to be propelled in three distinct modes, selected using a multi-position rotary switch mounted on the dashboard. The switch, shown in Figure 5, has four positions (in clockwise order); OFF, electric only mode (EM), hybrid mode (HEV), and combustion only mode (ICE). The OFF position is obvious. The others will be described.

ELECTRIC MODE - Electric only mode is the primary mode of operation for the vehicle. To engage, the driver must turn the ignition key to the ON position and the mode switch to EM position. The Suzuki engine computer, fuel injector, and

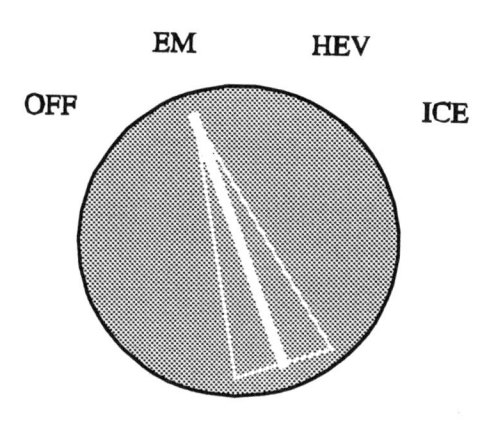

Figure 5.   UoA HEV Mode Select Switch

starter circuit are locked out in this ZEV operating mode. The electric motor output and hence the vehicle speed is controlled by the accelerator pedal. The accelerator pedal is linked to a potentiometer which sends an identical signal to each of the two motor controllers. The signal essentially demands torque, causing the controllers to supply more current to the motors and the motors to accelerate the vehicle. This electric motor control scheme is much simpler than some of the motor speed control systems contemplated. Beyond simplicity, it has the advantage that the electric drive behaves virtually identically to the normal gasoline engine drive. When changing gears, the clutch pedal and accelerator pedal are operated as in any other manual transmission vehicle.

It should be noted that the electric motors do not idle - no sound is heard from the engine bay when the vehicle is stopped. A low electrical whine (quieter than the normal combustion rumble) is noticeable when the vehicle is in motion and there is some belt noise from the toothed belt which is relatively poorly shielded on this prototype vehicle.

HYBRID MODE - The hybrid mode of operation is useful for situations like overtaking or driving up steep grades. As well as the electric motor control pots, the accelerator pedal is connected to the combustion engine throttle and thus determines the output of that engine when it is running.

Hybrid mode can be engaged with the vehicle moving or stopped and from either electric mode or gasoline mode. In switching from electric mode, the operator will turn the selector switch from the EM position to the HEV position. The Suzuki engine computer, fuel injector, and starting circuit are now enabled and all of the electric motor systems remain active and operational. The operator has to turn the ignition key to the START position to start the internal combustion engine as usual. (It was felt that an auto-start feature would add complexity and failure modes while not significantly enhancing the vehicle. Drivers already know how to start their engine!) The combustion engine will not couple with the transmission until its output shaft speed matches the rotational speed of the clutch driven by the electric motors. Once the combustion engine has coupled with the transmission input shaft, the torque from both power plants is delivered to the transmission.

In switching from gasoline (ICE) mode, the operator turns the selector switch from the ICE position to the HEV position. The gasoline engine is already running and the electric motors are permanently coupled so the additive torque from the electric motors is instantaneous. Because the available torque from the electric and gasoline sources is very similar, the performance with both power plants running is dramatically improved. A comparison of predicted performance figures in different modes can be seen in Figure 4.

GASOLINE MODE - This mode of operation is intended for situations like long distance highway driving. The operator has the option of either starting the vehicle in this mode or switching to this mode while the vehicle is in motion. To start the vehicle in this mode, the operator sets the Selector Switch to ICE position. The gasoline engine systems are enabled and the accelerator potentiometer inputs to the electric motor controllers are disabled int his position. The operator starts the engine using the ignition key and drives off, just as with a conventional automobile.

To engage this mode while the vehicle is in motion (assuming that the car is currently operating in electric only mode), a short transition through the hybrid mode is recommended. This allows the internal combustion engine to be started without the loss of power while driving. Once the gasoline engine is engaged, a transition from hybrid to ICE mode can safely be made.

AUTOMATIC MODE SWITCHING -The mode switching process could easily be automated in the future, eliminating many of the manual controls found in this vehicle and relieving the driver of additional operational distractions. This automation could also be arranged to enhance the vehicle's efficiency and battery life by strategically switching modes at predetermined, optimum times. At this stage of development, there is inadequate information to determine what control algorithms would be used to control automatic mode switching so this has not been fully implemented on the prototype vehicle.

BRAKING STRATEGY AND ON-BOARD CHARGING - Regenerative braking is available in all three operating modes with the U of A HEV and uses the energy normally wasted in braking to recharge the battery. The electric motors are automatically switched to generators when the driver presses the brake pedal, with the first portion of the pedal travel being electrical braking and the last portion applying the hydraulic brakes. This allows for the majority of braking energy to be reclaimed into the battery, while still maintaining the full braking capacity of the stock brakes. No regeneration occurs while the vehicle is in neutral or when the clutch pedal is depressed. This is to prevent possible damage to the motors as a result of them coming to a halt too rapidly, and thus throwing a magnet.

It should be noted that there is no engine braking from the internal combustion engine. With the U of A coupler in place, closing the throttle will allow the engine to drop to idle speed while the sprag clutch overruns. This is a change from normal vehicle operation but not a large one. It has the advantage of saving all the vehicle kinetic energy for the regenerative brake system.

In ICE mode, a separate regeneration control allows the internal combustion engine to be used to recharge the battery pack. The driver pushes a red push-button on the vehicle dashboard and adjusts a control knob to select the level of regeneration. With the prototype system, this is useful for recharging the batteries in areas where there are no suitable charging facilities. On a production vehicle, the capability could be especially useful in a situation where certain traffic zones are designated zero emission zones. When driving between these zones, this mode could be used to charge the battery pack and extend the electric range available. While this capability can be very useful in adding flexibility to the car, it must be emphasized that it is generally much more cost-efficient to use grid electrical power for battery charging than on-board power produced from highly taxed gasoline!

EMISSIONS CONTROL

In selecting an internal combustion engine, the desirable characteristics included state of the art emission control and the ability to easily alter the engine control process. The 1.0 litre Suzuki engine has electronic engine control and meets California emissions standards. However, it will typically operate at a higher power level in the HEV than it would in a Suzuki Swift. For example, the maximum power requirement during the limited speed hot-505 emissions test cycle specified for this competition is approximately 16 kW (22 Hp) in a 775 kg Swift and approximately 33 kW (44 Hp) in the 1575 kg HEV. This results in a need for some engine control changes and probably additional catalytic converter volume. The proposed engine control changes were to limit the high power enrichment and spark advance built into the Suzuki engine management system. These only come into effect when the engine is operating at high engine speeds and high torque, well beyond the normal operating range for a Swift running the FTP test cycles. The proposed changes push this enrichment closer to the extreme throttle position and engine speed limits. This leaves the maximum engine torque the same, but lowers emissions in what has become the new normal operating range.

Time constraints did not permit testing after recalibration of the engine controls so these changes were not made to the competition vehicle. In fact, the original, (small) catalytic converter was used and it failed due to overheating at the higher power level. Predictably, with enriched operation and a small catalyst, the U of A HEV did not meet current

Table 3
Emissions Standards and Competition
Performance of the U of A HEV

	NMOG (g/mi)	CO (g/mi)	NOx (g/mi)
Standard	0.32	4.4	0.7
U of A HEV	0.29	5.9	0.5
TLEV	0.16	4.4	0.7
LEV	0.10	4,4	0.4
ULEV	0.05	2.2	0.4

emissions standards in competition testing. Table 3 shows both the results obtained and the target emissions standards.

ENERGY CONSUMPTION

The Ford Escort station wagon has a published urban fuel economy rating of 6.73 litre/100 km with a curb weight of 1180 kg (2600 lb). The U of A hybrid conversion has increased the weight about 40% and this raises the tractive energy requirement due to increased inertial and rolling resistance. A prediction based on tractive energy analysis of an urban emissions test cycle shows that the energy requirement should rise by 29% due to the greater vehicle mass. The HEV would thus be expected to use about 29% more fuel when driving the same cycle as a (lighter) stock Escort wagon. However, testing showed that the 40% increase in vehicle weight was completely offset by the effect of lowering the maximum speed obtained during the urban driving cycle and of running a smaller engine at a higher, more efficient power level. Testing showed that the effect of the reduced speed alone allowed the heavier vehicle to match the standard fuel consumption (6.73 litre/100 km).

In competition, using the smaller Suzuki engine, the U of A vehicle had a fuel consumption of 6.78 litre/100 km over an urban/highway split. It might be assumed that the fuel efficiency would improve by using the smaller engine at a higher, more efficient output level. However, the combination of high power enrichment and higher operating speeds balanced this effect giving virtually standard fuel consumption. It is hoped that the planned engine control modifications will improve fuel economy in the future.

The electrical drive train efficiency of the U of A vehicle was 53.8% or 9.96 km/kW.hr as measured at the competition. This was on a driving cycle that was roughly comparable to a highway cycle (some 'stop & go' with a majority of 90 km/hr running). The electrical energy consumption during a highway cycle would be relatively unaffected by the energy recovery associated with regenerative braking. On a city cycle simulation, this should allow up to 15% reduction in electrical energy consumption. This feature is also available in hybrid and combustion-only modes so that a similar savings in total energy consumption (but no savings in fuel consumption) can also be expected in these modes. The actual regenerative efficiency of the components involved remains to be determined but using a relatively low regenerative braking level and typical battery charging efficiency, an 8% reduction in energy consumption should be attainable. One problem which became apparent during competition testing was that the current inputs during regenerative braking could drive the battery voltage above the peak safety levels for the controller. This was only a problem when the batteries were over approximately 60% state of charge, but it clearly requires some further sophistication in the controller.

From a marketing standpoint, the consumer wants a comparison between the operating economy of the gasoline engine and the electric motors. The cost of gasoline in Edmonton is Cdn$0.45/litre ($1.90/Imp.gallon) and the retail cost of electricity is approximately Cdn$0.07/kW-hr. The range of the U of A HEV, on battery power alone, is approximately 72 km (45 mi.) and the battery pack energy capacity is approximately 10 kW-hr. The fuel cost to travel 72 km in the HEV on gasoline power, at 7 L/100 km would be about Cdn$2.25. This same distance travelled on electric power alone would have a "fuel" cost about $0.70 CAN (35% of gasoline cost). However, because of the considerable cost and limited life of the battery, combined with limited daily mileage available in electric mode, the actual economics would have to include some charge for battery amortization. Since practical battery life has not been determined and our own battery costs are (hopefully) unrealistic for future production vehicles, a cost comparison including battery amortization has not been attempted.

VEHICLE STRUCTURE MODIFICATIONS

The U of A HEV is one of eighteen Ford Escort conversion vehicles in this competition. During the conversion process, some structural modifications were made to the Escort chassis in the engine bay and the passenger compartment. The students' execution of these modifications was critical to success of the car.

ENGINE BAY - The power train mounting system is one of the most important parts of the U of A's HEV conversion. Two electric motors, their controllers, the Suzuki engine, and a variety of digital monitoring equipment all needed to be integrated within the confines of an engine bay originally designed to house the Ford 1.9 L engine. The complex geometry of this new power plant combination and the need for precise alignment between each of the power train components required that the mounts be carefully designed and constructed.

The configuration and geometry of the various engine bay components was determined using dimensionally correct constructs of the Suzuki engine, the Ford transaxle, the electric

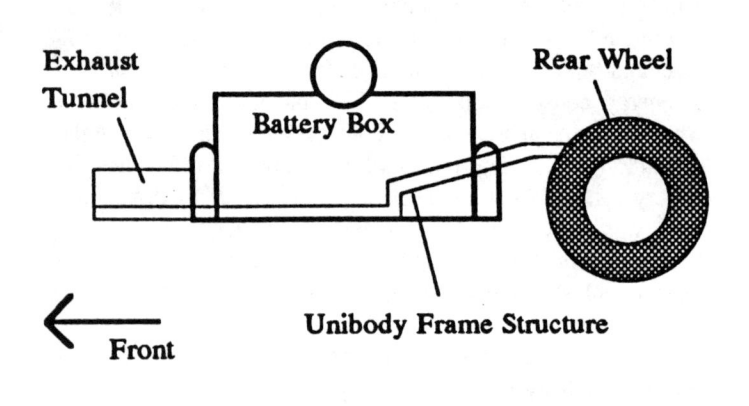

Figure 6. Side View of Battery Box Location Replacing Rear Seat.

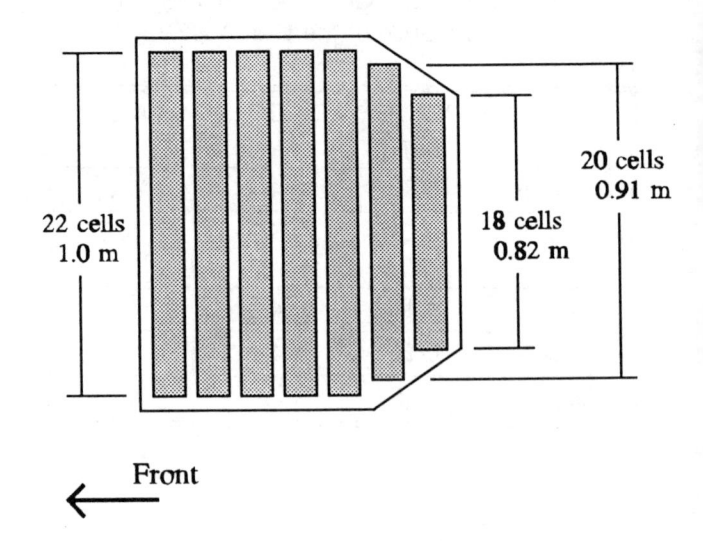

Figure 7. Top View of Battery Box Showing Cell Layout in Seven Trays (Rows) of Cells

motors, and the controllers. The two electric motors are mounted to a plate mounted to the transmission bell housing. This is mounted to the end of the engine with aluminum spacers to accommodate the thickness of the special coupler. Once these geometrical constraints were fixed, a detailed loading and stress analysis of the mounting plates and spacers was performed using Algor, a computer program for stress analysis. The main mounting plate was made of 6.35 mm (1/4 ") high strength steel. This plate weighs approximately 10 kg (22 lb) but it was felt a conservative design was essential for this critical component, particularly on the competition prototype. Mounting all the motors, engine and transmission on one central plate allows the entire power train to be installed as one operation similar to the stock vehicle.

Because of the limitations of the unibody construction, the original power train mounting points were used for the new power train even though they were not optimally arranged. Auxiliary mounts which extend from the various power train mounts to the vehicle frame were manufactured from aluminum to save weight. An important note is that the aggregate mass of the new power train including electric motors and controllers is very similar to that of the original 1.9 L engine and transmission. Mounts of similar stiffness were used for the new power train. However, testing has shown that softer vibration isolators would be desirable with the vibration from the three cylinder engine.

There are also a number of miscellaneous small mounts manufactured from 16-20 gauge mild steel which support peripheral engine bay equipment such as motor controllers and control potentiometers.

PASSENGER COMPARTMENT - The major modifications in this area were a box to accommodate the battery and a roll cage to meet competition safety requirements. The battery and its enclosure weigh approximately 318 kg (700 lb) and take the place of the rear passenger seat. This low, central location was chosen to optimize vehicle dynamic performance while retaining an essentially van-like vehicle capacity of two passengers plus a rear load area.

The floor pan in the centre section of the chassis was removed after a series of extensive strain gauge measurements and a complete finite element analysis, of the affected area using ANSYS software. This analysis showed that if the side rails and rear cross member were unaltered, the vehicle floor could be removed without compromising the structural integrity of the vehicle. In fact, the floor pan was replaced with the battery box but the analysis showed that the vehicle would not be placing undesirable loads on the battery box. Two schematic views of the battery box placement can be seen in Figures 6 and 7.

The exterior of the box is manufactured from 20 gauge steel closely matching the grade and gauge of the original Escort floor pan. The ideal material for the containment box is stainless steel which ensures corrosion resistance, however the cost was prohibitive. To provide increased corrosion resistance, the interior of the box is coated with a thick layer of water resistant epoxy and the exposed underside of the box is painted with a protective layer of undercoating. The battery box is attached to the vehicle with continuous seam welds running the full length of all sides. These welds are predicted to withstand a 15g deceleration of the vehicle in the event of a collision.

The battery box was designed to house 148 Nickel Cadmium cells in 7 rows as shown in Figure 7. Each row was supported on a removable tray which was slotted into the side walls of the box. These trays rest on two longitudinal rails along each side of the box. The row alignment was maintained by nonconducting links placed between rows. The battery was cooled by drawing cool air from below the vehicle and exhausting the heated air out through a vent located in the centre top of the box lid. Two fans drew air from a vent across the top of the box and exhausted it through ducts in the rear passenger doors. contain The box was completely sealed to the vehicle interior to prevent odour or other problems and the seals were located in areas with sufficient integrity to withstand collision deformation.

Important structures within the battery enclosure itself are the battery trays. Stainless steel was used for the construction of these trays. The need for these trays to remain undamaged by potential battery fluid spills offset the higher cost associated with using stainless steel.

Additional Modifications - Additional modifications to the passenger compartment were made for increased safety. Since the battery box was permanently mounted to the vehicle and may crush the vehicle in the event of a rollover, a rollbar was installed to increase occupant safety. Two styles of roll bar bracing were investigated; cross bracing and parallel bracing. Cross bracing is generally the most effective. However, with the U of A HEV design, finite element analysis showed that parallel bracing was adequate. As with standard roll bar arrangements, the main hoop would support the roof of the vehicle in the event of a rollover. For additional structural stability and as a safety measure for the occupants, a parallel brace was attached to the main hoop just forward of the battery containment box. This brace would protect the driver and passenger from the mass of the battery even in a severe impact where the battery box broke its welds. The main hoop was braced against the rear suspension mounts (aft of the battery box) to provide stability and increased structural stiffness for the whole vehicle.

Two racing seats with new mounts have been installed to provide higher occupant protection and to accommodate five point racing harnesses required for the competition. The mounts for the new racing seats are manufactured from aluminum extrusions and bolt directly to the vehicle floor pan using the original Ford mounting points. These seats have been placed as far rearward as the rollbar bracing has allowed.

SUSPENSION AND BRAKE MODIFICATIONS

The modified Escort designed and built by the U of A HEV project has a curb weight of approximately 1406 kg (3100 lb) which is 272 kg (600 lb) more than the stock Escort. The weight distribution has changed from a 59%/41% split to an approximate 55%/45% split on the front and rear axles respectively. The vertical centre of gravity of the vehicle has been maintained close to that of the unconverted Escort. The increased weight of U of A HEV prompted some modifications to maintain the original ground clearance and to increase spring stiffness to account for greater mass. The front spring coefficient was increased from 26.6 kN/m to 29.3 kN/m while the rear spring coefficient was increased from 15.6 kN/m to 18.8 kN/m.

Since the U of A HEV can operate as a fully electric vehicle, it will not always have engine vacuum to operate the standard vacuum-boosted power brakes. Rather than add another failure mode (i.e. a vacuum pump), a direct hydraulic system was designed. A performance test on the stock vacuum boost assembly showed that the unit could be replaced with a direct hydraulic system by changing the master cylinder size and the brake lever length. The stock 22.2 mm diameter master cylinder was replaced with a 19.1 mm unit. The changes resulted in maximum brake performance being achieved with a pedal force of 373 N which falls within the regulated range of 67 to 535 N as specified in SAE J937b (Service Brake System Performance Requirements - Passenger Car) [6]. The mechanical advantage of the brake pedal lever was changed from 4.12/1 to 4.73/1 by modifying the brake pedal mounting bracket, the master cylinder mounting bracket and the master cylinder pushrod.

With the increase in rear axle weight, consideration was given to replacing the rear drum brakes with disk brakes and/or changing the front/rear proportioning valve. Stock components were identified for such conversions but it was felt that the additional cost was not justified by any deficiencies in the braking performance. Since the U of A HEV was within the original car's GVW limit, no additional changes to the brake system were required.

BODY STYLING & ERGONOMICS

Body styling and ergonomics played a large role in the overall design of the U of A HEV. In every design decision, consideration was given to whether a component or assembly was aesthetically and practically located. This was recognized at the competition and the U of A HEV was given the Best Ergonomics Award. Some areas of the car were fundamentally unchangeable like the exterior styling of the Escort itself and the electric power disconnect switch. Other areas received a great deal of attention and detail like the vehicle dashboard, and the exhaust venting of the battery box.

VEHICLE DASHBOARD - A great deal of instrumentation and control mechanisms unique to an electric vehicle had to be incorporated into the U of A HEV. These include: a mode select switch, regeneration controls, a current meter, and a driver information system. The new dashboard, incorporating the additional instrumentation, was integrated into the U of A's modified Escort. It is an aesthetically pleasing and ergonomically practical design with flowing contours and strategic placement of all instruments and controls to prevent extra driver workload. An instrument array with a complete set of combustion and electric vehicle displays is included. Careful integration of the additional instrumentation lets the driver concentrate on the operation of the vehicle instead of searching for an instrument. Figure 8 gives a general schematic of the dashboard and instrument panel.

All of the driver-operated controls necessary to switch modes and control vehicle operating modes are conveniently

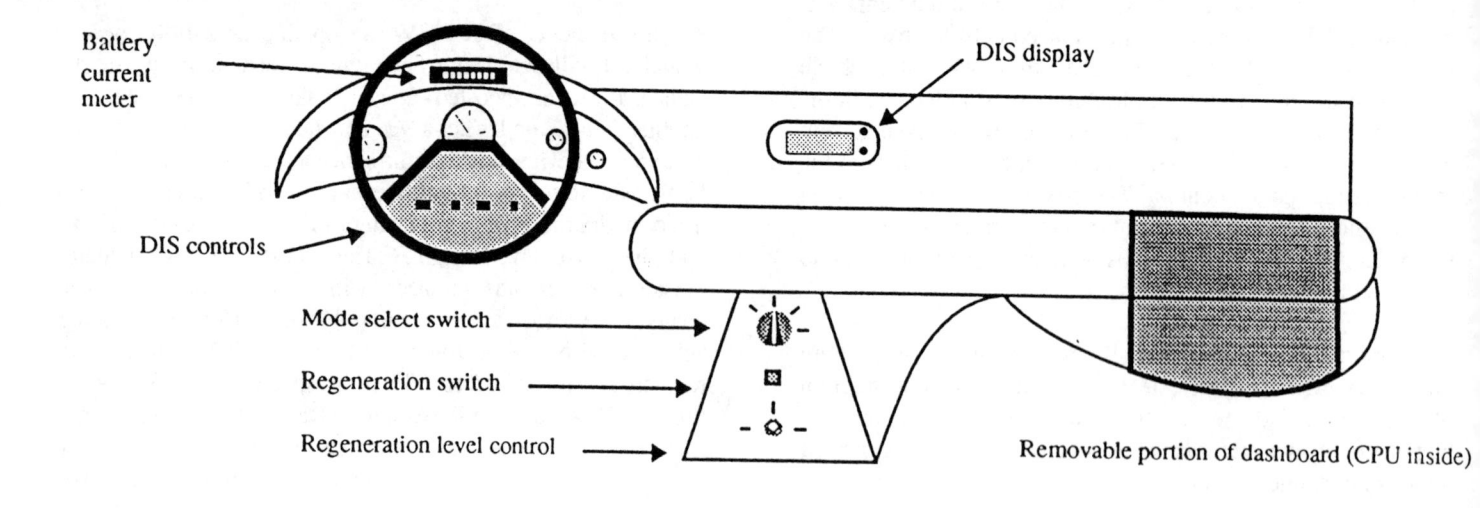

Figure 8. Schematic of the Prototype Dashboard Showing the Instruments and Controls Unique to the U of A HEV

located in the centre console adjacent to the driver's right hand. To change the operation mode (i.e. electric or hybrid), the driver will use the Select Switch which is located directly below the vehicle climate controls. This selector is a rotary position switch with four distinct positions; OFF, ELECTRIC, HYBRID, and GASOLINE. Also found on this centre console is a push button control and an accompanying potentiometer to control the level of regenerative charging in gasoline mode. This regeneration is engaged by depressing a red button which subsequently lights up to indicate its operation. The accompanying potentiometer uses a continuous rotary motion to select the desired level of regeneration.

Considerable thought went into the question of how much additional information to provide the driver given the multiple operating modes of this car. It was decided that a well-designed production car would need minimal additional instrumentation but the prototype would require a considerable amount of high precision instrumentation for testing purposes. In a production car, the only instruments required would be a battery charge indicator and a current draw indicator. Additional alarms should be available for battery temperature or other hazardous situations. There is no accepted method of determining state of charge for a large automotive battery so any instrument developed for this purpose inherently falls into the prototype category. As a solution to these conflicting requirements, the U of A HEV adds only two instruments to the dashboard. One is a simple current meter and the other is a computer-driven, multi-functional display with alarms.

The current meter is situated just above the standard instrument cluster. This meter uses a series of coloured LEDs similar to those used to indicate the level on a stereo. As the level of current being drawn from the batteries increases, the number of LEDs are lit proportionally. This provides the operator with a visual indication of how hard the electric motors are working, thereby reducing the risk that they or the battery will overheat. With maximum output only available for one minute out of three, it is important that real time information is easily monitored. This current meter is not particularly precise but gives the driver a good feel for relative power draw.

A more precise measurement system which has applications in the consumer market but was primarily implemented for use in the competition is the Driver Information System (DIS). This system monitors and displays real time information on a variety of critical vehicle subsystems including battery voltage and current, electric motor and combustion engine speeds, and the temperatures of the battery, motors, controllers, and engine. The information can be stored in memory and downloaded to an offboard computer for more detailed analysis. However, as a normal function, it is displayed on an LCD screen located in the top centre of the dash where it is easily monitored by either the driver or the front seat passenger. The DIS updates and displays this information to the LCD once per second. Since the volume of information is too large to be output in one screen, the driver has the option of scrolling through a variety of screens in order to obtain the desired information. The controls for the DIS are located on the vehicle steering column utilizing standard cruise control buttons. The DIS has the computer capability to perform various state of charge calculations and is being used to determine which is the most accurate. Additionally, the DIS is equipped with warning lights which will flash and automatically replace the existing display with any information which must be acted upon quickly. Examples of these interruptions include the battery or a motor overheating, or a terminal of the battery system shorting to chassis.

An additional feature which increases the functionality of the prototype dashboard is a removable portion directly in front of the passenger seat. This portion is directly above the glove compartment and is used to hold the master CPU. This allows access to the computer permitting easier modifications and repairs in the prototype vehicle.

EXHAUST VENTING OF THE BATTERY BOX - The air intake vents for the battery thermal management system are located in the floor of the vehicle underneath the battery box and are therefore hidden from view. The exhaust vent and its related ducting is attached to the top of this box. A small rectangular hole is cut into the bottom of each of the rear passenger doors to exhaust this air. From the exterior of the vehicle, none of this venting is obvious, and the vehicle aesthetics have not been compromised.

UNDER-HOOD VENTING - There was great concern about overheating motors and controllers during development. Some projections showed this could be a serious problem so a hood ventilation opening, with an attractively louvred plastic cover was added to the centre rear part of the hood. At this stage, it is still not clear that was required. It appears that the high efficiency of the electrical components chosen, combined with the low duty cycle in normal driving (and the short driving period at high power) prevents overheating with this power train. However, there have not been adequate tests in hot conditions to confirm this.

SUMMARY AND CONCLUSION

Eighteen months of imagination, design, creation, fabrication, and testing by University of Alberta students have resulted in a vehicle which was the best conversion vehicle and best overall vehicle at the 1993 HEV Challenge. The University of Alberta HEV runs on electricity, gasoline, or a combination of the two and would satisfy a market that wants an environmentally friendly form of transportation but won't sacrifice the comfort and freedom offered by today's automobiles.

Electric power is supplied by a nickel cadmium battery pack which can be charged overnight from a 220 Volt outlet. The combustion engine uses commonly available regular unleaded gasoline. The vehicle has a range of approximately 72 km (45 mi.) on electric power which would satisfy the needs of 77% of daily North American automobile travel. If additional range is needed, the use of gasoline gives a total range of over 500 km (311 mi.) without stopping. The performance is similar and adequate to keep up with traffic in either electric or gasoline mode. When added performance is required for passing or hill climbing, hybrid mode gives adequate power for spirited driving. The fuel cost to the consumer is predicted to be significantly less on electricity than gasoline so it should encourage users to travel in electric mode as much as possible.

The U of A vehicle, EMISSION IMPOSSIBLE, drives very much like the standard Escort. All of the common controls, instrumentation, and features of conventional automobiles are found in its design and the additional operating capabilities have been incorporated without significant complication for the driver. Additionally, this vehicle has maintained all of the standard occupant safety features and is structurally secure in the event of a collision.

At the competition, this vehicle produced emissions which do not quite meet current standards when operated in purely gasoline mode. However, emissions will be significantly improved just by reprogramming the fuel and spark control algorithms in the Suzuki engine management system.

The U of A HEV was designed to be put into production as a practical and marketable automobile. Wherever possible, the modifications made to this Escort have incorporated ideas and methods that will lead to simple manufacture at a production level. The majority of the components used in the conversion are readily available and currently being produced. The more exotic and expensive components including the electric motors and batteries could become commonplace with increased demand. The market appeal of this vehicle is that it looks and drives much like a conventional automobile with all of the usual comforts and functional aspects retained. It has stock production headlights, horn, steering wheel, accelerator pedal, brake pedal, and shifter, all in their stock locations, allowing the operator to feel comfortable when driving. The additional controls and instruments have been kept to a minimum and have been added in a way which does not increase the driver's workload.

The electric technology of the future has been successfully integrated into an automobile for today. The U.S. Department of Energy, SAE, and Ford motor company have tapped into the resources and potential available within the North American University and College system to find a solution to a pressing environmental problem: air pollution caused by automobiles. The hybrid concept may not be the final solution to this problem but it is a necessary step in the right direction and will help ease the public into an emissions free automotive future.

REFERENCES

1. "California Air Quality, A Status Report", California Air Resources Board, Sacramento, 1991

2. M.D. Checkel, "NGV Conversion Effects on Vehicle Emissions and Fuel Economy, Appendix A 'Development of Multi-mode Test Schedule' ", Report for Alberta Transportation and Utilities, R & D Division, January, 1992

3. "Emission Test Driving Schedules", SAE Information Report, SAE J1807, SAE Handbook

4. G Sovran and M.S. Bohn, "Formulae for the Tractive Energy Requirements of Vehicles Driving the EPA Schedules", SAE paper 810184, Annual Congress, Detroit, 1981

5. P. Wasielewski, L. Evans and M.-F. Chang, "Automobile Braking Energy, Acceleration and Speed in City Traffic", SAE paper 800795, Passenger Car Meeting, Dearborn, 1981

6. "Service Brake System Performance Requirements - Passenger Car", SAE Standard J937b, SAE Handbook

940340

Hybrid Electric Vehicle Development at the University of California, Davis: The Design of Ground FX

Rebecca Riley, Mark Duvall, Robert Cobene II,
Gregory Eng, Keith Kruetzfeldt, and Andrew A. Frank, Advisor
University of California, Davis

Abstract

The last few years have been an exciting time for alternative vehicle development. New concerns about the environmental impact of personal transportation and about the United States' dependence on imported oil have pushed energy efficient, ultra-low, and zero emissions vehicles to the forefront of automotive design. California's own mandate for Zero Emissions Vehicles (ZEV) takes effect in 1998, creating a tremendous push towards the difficult goal of producing a commercially viable, practical electric vehicle for sale in 1998. Beyond California, most of the world's automakers are simultaneously committing tremendous research and development resources towards the technology necessary for a viable electric vehicle.

The University of California at Davis is one of seven California universities participating in the 1993 Ford Hybrid Electric Vehicle Challenge. The Vehicle Design Team in our College of Engineering is one of thirty in the United States and Canada, each developing its own brand of viable two passenger hybrid electric automobile for this competition. The sponsors of this competition, Ford Motor Co., the Department of Energy, and the Society of Automotive Engineers, have established common guidelines and performance goals for the schools to attain.

UC Davis Hybrid Vehicle Design Concept

Among the benchmarks for the HEV Challenge are a range of 40 miles at 40 mph on electric power alone (ZEV Mode), a range of 200 miles using both electric and internal combustion (IC) power (HEV Mode) and Transitional Low Emission Vehicle (TLEV) rating. Our design team combined these goals with our own performance guidelines. This allows us to establish a set of standards for our vehicle that are appropriate for the Challenge *and* reflect the anticipated needs of drivers of the area we live in. We arrived at the following objectives.

HEV Design Goals

1. 60 mile range at 60 mph in ZEV mode.
2. 100 mpg fuel economy at 60 mph with IC engine *only*.
3. 0-60 mph acceleration in under 10 seconds in both ZEV and HEV modes.
4. Unlimited range in HEV mode.
5. TLEV emissions rating with IC engine.
6. Safety and driveability comparable to a conventional auto.

These goals are very demanding, but it is our desire to use this vehicle to prove to the consumer and to our own state government that the hybrid electric vehicle concept can produce not only a superior electric-powered vehicle, but also a superior automobile. To the best of our knowledge no single car, prototype or otherwise, offers this combination of performance, zero-emissions capability, and fuel economy. That fact alone makes these goals worth realizing. It is our hope at the university that this car can serve as the stepping stone from the conventional automobile to the electric, hybrid electric, or other alternative cars of the future.

But first, we must prove to the country's drivers that we can provide a car with superior emissions and energy efficiency without sacrificing the convenience and performance to which they are accustomed.

The Vehicle Design Team's goals helped to form the concept. To attain these objectives, the vehicle's design must concentrate on aerodynamic body design, light weight, and an innovative powertrain capable of meeting the performance criteria.

The design and construction process began in June, 1992 and will be finished for the HEV Challenge scheduled for June 1, 1993 in Dearborn, Michigan. The UC Davis hybrid electric vehicle is a two passenger commuter sports car. The finished vehicle weighs approximately 1800 lbs. The fiberglass body weighs under 160 lbs. including gull-wing doors and windows. The chassis is an aluminum space-frame with four-wheel independent suspension and a mid-engine, rear drive powertrain. Final weight for the rolling chassis is only 400 lbs., including tires, suspension, seats, and safety equipment.

The vehicle's powertrain features a similar level of creativity as the body/chassis. The integrated parallel hybrid electric drivetrain couples a 32 kW UNIQ brushless DC motor with a 570 cc, four-cycle Briggs & Stratton OHV motor with modern engine management and emission controls. These two powerplants mate through a common input shaft to a 5-speed transaxle. The battery pack is a 170 volt, 7.1 kWh, flooded Nickel/Cadmium array of 260 individual cells (two parallel strings of 130 each). The powertrain develops almost 95% of its peak HEV input torque of over 130 ft-lbs from 1500 - 4500 rpm.

The body of this paper explains both the Body/Chassis and Powertrain development for the UC Davis Hybrid Electric Vehicle. The HEV Challenge in June, 1993 will determine whether the vehicle meets its objectives.

Body and Chassis Platform Development

Vehicle Platform Design

The first step in the design process is an energy balance of the vehicle's power requirements. We commonly identify aerodynamic drag and rolling resistance as the two primary design dependent variables in efficient vehicle design. The limited electrical energy available in the battery pack mandates that the vehicle's design emphasize efficiency. The force equation is

$$Force_{drag} = \underbrace{1/2 \, \rho \, A_f C_d V^2}_{Aero\ Drag} + \underbrace{C_{rr} * W}_{Rolling\ Drag}$$

ρ : *air density*
A_f : *Maximum Vehicle Frontal Area*
C_d : *Vehicle Drag Coefficient*
V : *Vehicle Velocity*

C_{rr} : *Coefficient of Rolling Resistance*
W : *Total Vehicle Weight*

Bearing drag, which is usually minimal, is neglected here for simplicity.

Before lengthy simulations take place on a computer, hand calculations are useful to determine the capabilities of the vehicle design. We used several varieties of a best case/worst case design scenario. We estimated various vehicle parameters that affect drag and powertrain efficiency, and then came up with a low and high value. From these parameters, possible fuel economy and electric range figures were obtained. For the powertrain, estimates of electric motor and powertrain efficiency were developed. Also, the minimum average brake specific fuel consumption (lbs.fuel/horsepower*hour) for the APU can be used to generate a fuel economy prediction. The calculations are done for a steady-state speed of 60 mph. Later on, simulations will be used based on the Hot 505 cycle Ford has proposed for emissions testing.

As is evident from Table 1, the vehicle's drag can roughly double from the best case situation if the vehicle fails to meet design criteria. Most of the parameters used are less than 25% apart, yet the sum of the "overruns" could cause an increase in total power consumption of almost 100%. We must also remember that numbers and simulations are just that, and that the real world vehicle could possibly exceed the "worst case".

Table 1: Vehicle Design Parameters

Vehicle Parameter	Best Case	Worst Case
Frontal Area	16 ft^2	18 ft^2
Drag Coefficient	0.15	0.24
Loaded Weight	1900 lbs	2200 lbs
C_{rr}	0.006	0.008
Elec Motor Eff.	85%	75%
Drivetrain Eff.	95%	85%
BSFC minimum	0.50 lb/hphr	0.70 lb/hphr
Drag Forces		
F_{aero} at 60 mph	23.32 lbf	41.98 lbf
F_{roll} at 60 mph	11.4 lbf	22.0 lbf
Power Required		
P_{aero} at 60 mph	3.73 hp	6.72 hp
P_{roll} at 60 mph	1.82 hp	3.52 hp
P_{total} at 60 mph	5.55 hp	10.24 hp
ZEV Mode		
$P_{rear\ wheel}$	6.87 hp (5.12 kW)	12.68 hp (9.46 kW)
Watt-hours/mile	85.3	157.6
Range at 7.1 kWh	83.8 miles	45.3 miles
HEV Mode		
$P_{rear\ wheel}$	6.34 hp	11.28 hp
Fuel Economy at minimum BSFC	113.6 mpg	45.6 mpg

Table 2: Body Design Considerations

1. Minimum Drag Coefficient (C_D)
2. Small Frontal Area
3. Enclosed Wheels
4. Enclosed Powertrain

Aerodynamics

To achieve an extremely low drag coefficient the body should be shaped to maintain laminar flow as far aft as possible and have the smallest region of separated flow possible. We accomplished this by basing the initial streamlined body design on a NACA 66-025 airfoil.

The 6-series airfoil sections were designed to maintain laminar flow over more of the airfoil than older designs. Within this family, we chose a minimum pressure point location at 60% of the chord length (66-series). The airflow sees decreasing pressure over 60% of the airfoil, reaches a minimum, then recovers pressure. The large initial favorable pressure gradient increases the transition Reynolds number and permits laminar flow to exist at higher velocity and/or greater body length.[1]

Laminar flow is highly desirable because it results in much lower skin-friction drag than turbulent flow. Traveling at 60 mph, our car has a Reynolds number of 7.6 million. For smooth-surface laminar flow the skin-friction coefficient is 0.0005; turbulent flow has a skin-fricton coefficient of 0.0032. Even with pockets of turbulence, maximizing laminar flow dramatically decreases drag. [2]

Aerodynamically, a 67-series airfoil would perform even better with 10% more laminar flow. Practically, however, the different thickness distribution would not allow enough passenger and engine compartment space. Also, the NACA 67-series airfoil tends to have more flow separation in the aft region of the body at lower Reynolds numbers.

The airfoil shape was altered slightly from the original symmetric NACA form to perform better in the ground effect and package our whole our design more effectively. Five percent camber was introduced by using a NACA mean camber line of 0.7. This helps eliminate the negative lift due to close proximity to the ground plane. Cambering slightly raises the lower surface farther off the ground, reducing the pressure gradients beneath the vehicle. This assists in delaying turbulent flow and airflow separation.[3] This shape also allows us to enclose the wheels more completely.

Body Design

Our project objective of building an extremely efficient commuter vehicle required that we minimize the aerodynamic losses of our vehicle. There was no reason to waste power overcoming large amounts of drag when we were able to design an innovative body based primarily on aerodynamic efficiency.

Many factors, influencing everything from turn radius to engine cooling, had to be considered when designing the body. The main concepts behind our design are shown in Table 2.

Additional camber, however, would cause problems enclosing the rear wheels and powertrain components. It was also important not to introduce too much camber because we did not want to generate positive lift. Maintaining zero lift is important for several reasons:

- Lift causes induced drag
- Positive lift will reduce traction, reducing vehicle stability.
- Negative lift increases the force on the road, increasing rolling drag.

Sizing

Aerodynamic analysis of streamlined vehicle shapes shows that for a given thickness aerodynamic drag is minimized by using a vertical thickness equal to approximately 25% of the length. For a fixed thickness, ratios much larger than this will result in the flow separating further forward on the body increasing form drag. If the ratio is much smaller, the increase in length will cause a larger gain in skin friction drag than drop in form drag.

We determined the dimensions of the car based on two needs:

- Minimum frontal area
- Passenger compartment space adequate for two people and regulation cargo space.

For fixed conditions, minimizing frontal area minimizes aerodynamic drag and, thus reduces vehicle power requirements. A maximum body thickness of 41 inches was chosen because it allows just enough passenger head room. Using the 25% thickness-to-length ratio led to a design length of 164 inches. A maximum width of 70 inches was reached based on the track width requirements fitting within a tapering body shape and interior compartment needs including side impact crush zones. The frontal area is 17.5 ft^2.

Other body styling choices were based on literature with data on streamline shapes and existing aerodynamically designed experimental vehicles. Books such as *Fluid Dynamic Drag* by S.F. Hoerner and *Impact of Aerodynamics on Vehicle Design* ed. M.A. Dorgham contain drag coefficients and aerodynamic characteristics of many bluff and streamline body shapes.

[4,5] Recent examples of experimental vehicles, including GM's Sunraycer and Ultralite, and our own Supermileage vehicles are excellent examples of aerodynamic efficiency. The Sunraycer is an extreme example of design for aerodynamics, showing that a 6-series airfoil body can be used effectively for a vehicle in the ground effect. It ended up with a drag coefficient of 0.095 to 0.108. [6] The Ultralite's design shows that a practical car can be designed around aerodynamic efficiency. It is a four passenger vehicle with a drag coefficient of only 0.192. [7]

From these resources we determined that tapering the planform shape from front to back and curving the sides of the vehicle would minimize drag. We kept the same airfoil throughout the width of the body, varying thickness (and, consequently, length) to taper body height and length. The amount and rate of variation was determined by the size of the passenger compartment and the wheel track width requirements. This shaping resulted in a gradual decrease in maximum width from 70 inches at the lower surface to 42.5 inches near the upper body surface, which should reduce the adverse drag and stability characteristics in crosswinds. From front to back the body tapers from 70 to 53 inches. See Figures 1 & 2 for our original and final body designs.

We did not have access to either large-scale wind tunnels or sophisticated computational aerodynamics codes such as VSAERO to analyze and optimize our design. Local CFD experts told us that codes cannot accurately perform a three-dimensional, viscous flow analysis over a complex body accounting for the ground effect and parasitic elements such as door handles and seams between body panels.

We did perform some two-dimensional analysis to estimate that the zero-lift line for our airfoil section was -2.5°. Thus, the body should be angled slightly nose-down to maintain zero lift. This was only an order-of-magnitude estimate to use as a starting point. For the real three-dimensional case the true zero-lift angle must be determined experimentally.

To satisfy the need to reduce aerodynamic drag caused by wheels, the tires are enclosed within the body as much as possible. Fairings below the level of the body are not used because they would require a large addition in frontal area to allow room for the wheels to turn, increasing aerodynamic drag. Full fairings also have the potential to prevent cross flow under the car causing a reduction in the downward lift force.[6] This would cause problems due to an effective increase in upward lift.

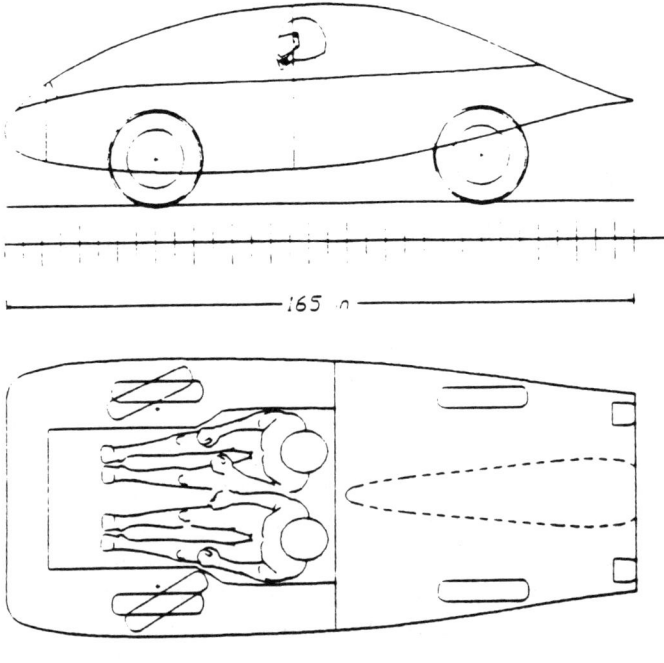

Fig 1. Original Body Design

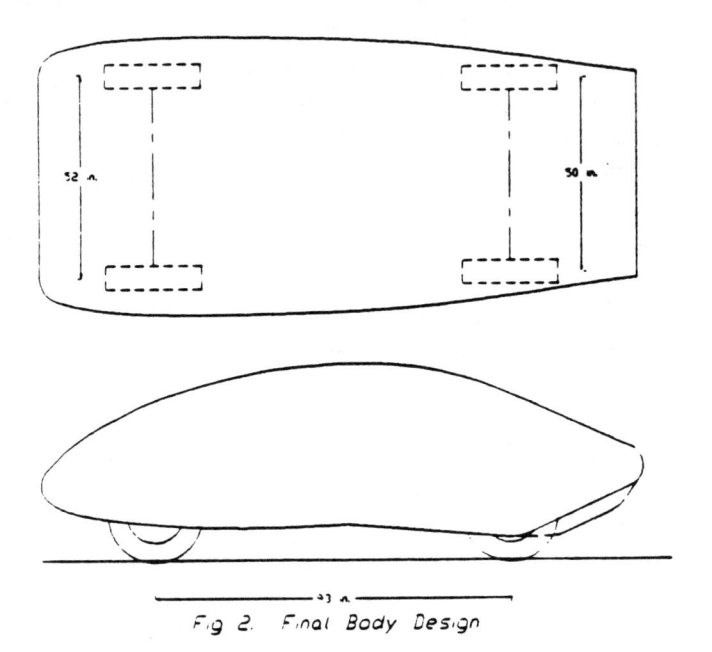

Fig 2. Final Body Design

Internal Airflow

Both the passenger and engine compartments need ventilation to supply airflow for cooling. Engine cooling systems cause drag which can increase the overall drag coefficient by 2% to 10% depending on how carefully the system is designed. The passenger cooling system does not have as large an effect on drag due its comparatively low air flow rate. [5]

In designing the cooling systems we placed intakes in high pressure areas and outlets in low pressure regions (relative to interior pressure). For 6-series airfoils pressure decreases as the air flows aft, so passenger and battery ventilation inlets are near the nose of the car where the pressure is highest.

Engine cooling inlets are placed on the underside of the body just before the engine compartment. Flow is diffused through the mechanical engine fan and exhausted at the location on the body underside with maximum curvature, as this is where minimum pressure occurs.

Redesign

Only one large-scale redesign was done to the body. The powertrain configuration and placement was not completed exactly as foreseen at the start of the project. Because of the long time needed to manufacture the body the shape was determined based on guesses of sizes and placement of all other components. Late in the manufacturing process we found that some powertrain components and frame members would not be concealed within the airfoil shaped bodywork.

From both an aerodynamics and an aesthetic point of view we didn't want to simply cut holes in the body and leave parts exposed in the airflow. At 60 mph the drag on the rough drivetrain and frame components could require an additional 1.5 hp based on a reference drag area of 1.4 ft^2. When the aerodynamic drag of the whole vehicle is overcome by only 4.64 hp, that is a significant requirement.

To remedy this situation we redesigned the underside and back end of the body. To create the additional space needed we changed the region of the airfoil which sloped upward beneath the engine compartment. Instead, we kept the lower surface closer to the ground 40 inches further aft. Slightly behind the rear wheels we curved the lower surface up a smoothly as possible around the transmission to meet the existing upper body surface.

For ease of manufacture and to minimize flow separation caused by a sharp curvature, the panel still did not enclose the transaxle completely. For aesthetics and aerodynamics we designed a small additional housing.

This underbody redesign provided several benefits, including:

- a reduction in aerodynamic losses due to form drag on transaxle, etc.
- additional space for cargo and other vehicle components.
- increased stability in crosswinds
- a convenient surface for taillights and license plate.

Manufacturing and Materials

The full-scale body plug was first carved out of Styrofoam using masonite stations every five inches to guarantee the correct shape. A layer of fiberglass and body filler was added and sanded completely smooth. Part lines were scribed onto the plug, dividing it into panels for wheel covers, doors, engine compartment access, etc.

Next, a fiberglass female mold was made for each panel. The final body panels were made using these molds to ensure that the outer surface of the car was smooth.

The body is made of a composite sandwich structure comprised of 6 oz./sq. yd. woven pre-impregnated fiberglass as faces on each side of Nomex honeycomb core. A typical body panel is 2 layers of fiberglass on each side of 0.250 in. Nomex honeycomb. Door panels required 6 layers of glass due to the large window areas. Wheel housings were stiffened by foam strips instead of honeycomb so that turn radius could be improved. For strength, additional plies of fiberglass were added in mounting locations.

One reason for choosing Fiberglass/Nomex was that we could easily form it to our unusual, nonlinear body shape. The primary reason, however, was the material system's high specific strength and stiffness. The sandwich averages 0.52 lb/ft^2, yielding a body shell with of 68 lbs, excluding doors and windshields. The thicker doors with hinges, locking mechanisms, etc. add 20 lbs. The front and rear windshields, made of 0.250 in. Lexan, add 60 lbs, bring the total body weight to 148 lbs.

The body panels are attached to one another using nut plates. The panels overlap slightly with the outer panel bolting to a nut plate riveted to the inner panel. The body panels are secured to the frame by bolting the panels to frame members.

Ergonomics

The goal of the interior design was to ensure that the human factors involved in vehicle design and use were not neglected. The passenger compartment must be safe, comfortable, and attractive. The driver must have good visibility of the gauges and landscape, and all controls must be within easy reach.

We wanted the interior to also reflect the style of the exterior body. Thus, the main interior element, the dashboard, is formed of a series of curves with few linear edges or flat surfaces. This style also allowed us to design a wrap-around instrument panel that created more space for gauges than a typical panel. Our instrument panel covers 105 sq in, 130% larger than the panel of a Ford Escort. This large size is necessary because we need additional gauges to monitor both the electric powertrain and the alternative power unit (APU). Even with the extended instrument panel, all gauges fall within a space that is easy for the driver to scan quickly.[8]

The upper surface of the dashboard starts just behind the front bumper and extends 40 in. back. The occupants sit an additional 30 in. back. Both seats are mounted at a 12° angle to provide additional headroom so that the helmeted passengers will meet the HEV Challenge eye-height and 2 in. rollbar clearance requirements. The seats recline as well, allowing occupants of different heights to adjust their position for best visibility and comfort. Additionally, the driver's seat slides 28 in. along the length of the body to allow drivers from 5'0" to 6'3" in height to easily reach the pedals and other controls.

Visibility

To maintain good visibility with our low-slung body design, we maximized window area and positioned rear and side view mirrors carefully. The front windshield extends from the top of the car to the bumper, and wraps around the sides of the body, covering 24 ft^2. The rear windshield covers 14 ft^2. This is ample, especially when the driver uses the rear view mirror to see behind the vehicle. The doors each contain 4 ft^2 of window, divided into side and roof windows. The upper windows allow the occupants to see objects above and in front of the car as well as lessening the feeling of being in a enclosed space. The side windows provide excellent visibility to both sides of the car.

When trying to minimize the body's parasitic drag, we experimented with interior mounted side mirrors. We found that for our design, mounting the mirrors near the A-pillars within the body does not reduce visibility when compared to conventional exterior mounted mirrors.

Passenger Comfort

Passenger cooling is accomplished by fresh air ventilation. Heating is provided by a resistance heater powered by the main battery and regulated by the vehicles computer controller to prevent excessive battery drain. The airflow is concentrated at the occupants' heads. Military studies have found that if the head is kept at a comfortable temperature, the rest of the body will typically feel more comfortable as well. Ducting is used to route air to the windshield for defrosting.

Another aspect of passenger comfort we considered is long-term exposure to noise. The overall noise level in the passenger compartment is the sum of the noises generated by the power plant, tires, and airflow. At 60 mph, wind noise above 1000 Hz can reach 68 dB(A). Our electric motor drive belt specifications say that it may reach 110 dB(A) at 3700 Hz. Because these worst-case levels are at high frequencies we looked for a material that would be effective in this range, yet not add much additional weight to the car or reduce space in the passenger compartment. We chose 0.50 in thick sheets of embossed polyurethane foam with a density of 2 lb/cu ft. It will absorb up to 95% of airborne sound in high frequency ranges.

Chassis Development

Frame

In the formulations of our vehicle, several ideas were considered for the structural chassis. A light weight chassis was a necessary facet of our vehicle design. We studied three basic chassis concept: composite monocoque (GM Ultralite), ladder frame, and tubular spaceframe. The idea of a composite monocoque similar to the GM Ultralite had been considered but rejected on the basis of cost and complexity. In addition, this is not current method of mass producing a vehicle. A separate frame and body shell has an additional development advantage, as the chassis can be finished and tested independently of the body. The more independent nature of the structure and the shell allows us to concentrate on weight reduction in the body without the vast structural concerns of the monocoque. The real-world result is that for a prototype of this nature, a frame and shell design is often lighter than a first-run composite monocoque.

Initially, hand modeling and Finite Element Analysis (FEA) were used to create frame concepts. The roll hoop and side bolster requirements of the HEV Challenge seemed to favor a tubular spaceframe constructed around these structural points. Repeated FEA simulations indicated that a frame constructed primarily of large diameter tubing would yield a light, rigid structure, while incorporating the roll and side protection features specified by Ford.

Several scenarios were considered for collision and as a result the frame was designed around a roll cage. Due to the light weight structure necessary for a low rolling resistance vehicle, the frame is constructed of an aluminum alloy. The aluminum tubes are welded together by the TIG welding process which produces pronounced strength reduction in the areas surrounding the welded joints, the Heat Affected Zone (HAZ). Unfortunately, the weight savings (aluminum is about 1/3 the weight of steel) of the aluminum over the steel is diminished by its lower fatigue resistance.

In design, one way to eliminate this fatigue problem is to thermally treat, solution heat treat and artificially age the whole frame. However, there was not a heat treating facility nearby that could heat treat our whole frame. Another solution to the fatigue problem is more fundamental and still solved our weight problem. We over-designed the structures to minimize deflections and, hence, stresses imparted on the aluminum joints in the frame. This approach provided a somewhat heavy frame but still a lower weight structure than steel.

A combination of past experience and the FEA results prompted the selection of large diameter aluminum alloy tubing as the primary frame material. We used 2.5 in. diameter, 0.065 in wall aluminum alloy tubing (6061 T6 alloy). Rectangular tubing of the same alloy was incorporated in specific engine mounting point. Alloy sheet metal serves for the floor pan and exterior battery enclosures. The final bare frame weighs only 150 lbs., contributing to an extremely light rolling chassis of 400 lbs.

Suspension and Brakes

We were well aware from the beginning that vehicle suspension systems, with associated handling characteristics, are very difficult to "design right" the first time. With this in mind we decided to use existing suspension technology that satisfied safety, strength, and light weight criterion. A steel frame prototype, or "mule", was used to develop the vehicle handling. The mule simulated the track, wheelbase, and weight of the final vehicle.

A MacPherson suspension system was chosen due to its simplicity and ease of modification. We decided to use the existing suspension from a Chevrolet Sprint passenger car for our rear system because it fit this criterion and is a driven system. The front system uses a Datsun 1200 because it is easily modified. These suspension components have proven to be very satisfactory on our prototype chassis. In fact, the handling characteristics of both systems on our prototype chassis out-performed most sports cars in a slalom course.

The development of the prototype chassis taught us that for our vehicle, the low center of gravity, lower weight, and mid-engine design would provide inherently better handling than the average conventional commuter vehicle. This enabled us to concentrate our energy on frame and drivetrain development, as we were very confident that our modified suspension components would work well. With our handling data from the prototype, we lengthened the wheelbase from 85 in. to 93 in. to improve front/rear weight distribution, increase passenger space, and smooth the vehicle's ride.

Again in the interest of a light weight structure, the brakes selected for our chassis are aluminum alloy disk brake components. These were purchased from a manufacturer of racing and specialty disk brake systems.

Tires and wheels

The initial calculations shown in Table 1 indicate that rolling resistance is about one-third of the total drag on the vehicle at 60 mph. This is a significant source of power loss. Aerodynamic vehicles always have a disproportionate amount of rolling drag due the dramatic reduction in aerodynamic drag. (P_{drag} for a Ford Festiva at 60 mph is over 12 hp vs our HEV estimate of about 6 hp.)

Tire selection is of utmost importance in a high efficiency vehicle. With this in mind our team researched tires available on the market as well as those not available on the market. Our first selection for our vehicle tires are the Firestone Tempa P195/90R14 space saver tires, a compact spare from a Ford Thunderbird. The engineers at Firestone quoted numbers up to 30% lower in rolling drag than tires that are commercially available. The Ford HEV rules and regulations panel disallowed the use of these tires in the competition on the grounds of the temporary nature of the tires, even after documentation diplaying the safety of prolonged use was submitted to them.

Further research revealed that the tires used on the Ford Ecostar produced equivalent rolling resistance to the unuseable space saver tires. These Firestone P 195/70R14 EVT (Electric Vehicle Tire) have a C_{rr} of only 0.006, meeting our best case estimate for rolling drag. Unfortunately, these tires demanded more space and our chassis and body design were modified accordingly. However, due to the fact that these electric vehicle tires are not DOT approved, the Ford HEV regulations board once again disallowed the use of these tires on our vehicle. This forced our team to select an alternate tire and decrease the efficiency of our vehicle.

With these two setbacks, we decided to use the Goodyear Invicta P195/70R14 tires which are also being used by our competition. These tires are still excellent low rolling resistance tires and with the help of Goodyear engineers we selected adequate tires and inflation pressures for our vehicles size and weight distribution. Figure 3 illustrates the importance of low rolling resistance in a comparison of these designs with a standard tire [9]

Fig. 3

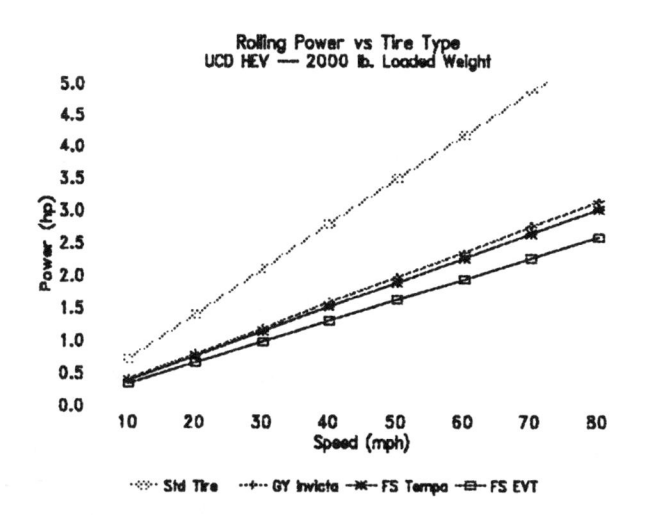

Rolling Power vs Tire Type
UCD HEV — 2000 lb. Loaded Weight

[Graph: Power (hp) vs Speed (mph), with curves for Std Tire, GY Invicta, FS Tempo, FS EVT]

Battery Enclosure Safety

Our vehicle uses a 7.1 kWh battery pack weighing 525 lbs and occupying over 4 ft^3 of volume. It is difficult to safely place this much volume and mass, as the shape of our aerodynamic body and the enclosed nature of the vehicles wheels created unusual space enclosures within the vehicle. We ruled out the center packaging used in the GM Impact due to an unacceptable addition of width (12") and the corresponding increase in frontal area. The center tunnel will also be difficult to access. Another option, front packaging, places too much weight outside the confines of the wheel base, and leads to extended battery cables.

The enclosed space between the front and rear wheel on each side presented an ideal place to locate the battery enclosures. The HEV Challenge regulations already mandate a side bolster to protect each passenger. The frame design was modified to include an integral battery box on the right and left sides of the vehicle, between the front and rear tires. Each box is protect by four aluminum alloy 6061 T6 side impact members, two 2.5 in. OD 0.065 wall tubes, and two 1" x 2" 0.125 wall rectangular members.

To further protect the batteries, the individual cells are enclosed in rigid composite boxes and separated by nylon insulators. The interior of each aluminum box frame is electrically insulated with a combination of 0.125" nylon sheet and 0.125" neoprene. Each box is accessible through the top and side of the frame. The enclosures are tightly sealed from the passenger compartment by a firewall. Ventilation is provided by one 41 cfm brushless DC fan on each box.

Battery safety is obviously a prime concern in the development of an electric-powered vehicle. The very nature of our cells improves their safety. The individual polyurethane casings are extremely rigid. In the event of a side impact, the batteries are protected by four side members, the composite boxes, and their own rigid cases. Also the electrolyte, potassium hydroxide, does not present a caustic burn danger.

Manufacturability and Recyclability

Vehicle manufacturability was an influential factor in our decision to use a spaceframe and body shell construction. Our fiberglass body panels could be duplicated easily in production by plastic pieces. An optimized form of the aluminum spaceframe could take advantage of custom extrusions and possibly bonded joints. These are not new concepts to the automotive world: the Ford Contour features a bonded aluminum frame.

Light weight construction will require new manufacturing techniques. Our finished vehicle, in its first iteration, weighs only 1800 lbs, less than a Ford Festiva. This includes our 525 lb. battery array.

Our powertrain uses almost entirely stock components. Our engine modification are purely design differences. An optimized, mass-produced drivetrain, including an 8-valve, liquid-cooled v-twin APU could weigh almost 100 lbs. less and feature more APU power and better efficiency.

Recyclability is inherently strong in a vehicle that contains over 300 lbs. of aluminum alloy. The NiCad batteries, as well, are now completely recyclable at the new SAFT-NIFE Nickel/Cadmium reclamation plant in Europe (SAFT covers shipping costs).

Powertrain Development

Our powertrain design is the second part of our hybrid electric design concept. The vehicle platform we have developed will determine the energy and performance requirements of the powertrain. The powertrain must balance efficiency and power in order to meet our vehicle's performance goals.

To repeat, these goals are:

1. 60 mile range at 60 mph in ZEV mode.
2. 100 mpg fuel economy at 60 mph with IC engine *only*.
3. 0-60 mph acceleration in under 10 seconds in both ZEV and HEV modes.
4. Unlimited Range in HEV mode.
5. TLEV emissions rating with IC engine.

To meet these goals, we chose a rarely used HEV powertrain platform: the small APU/large electric motor parallel hybrid. The two powerplants are packaged into an integrated mid-engine, rear drive drivetrain. We selected this configuration for its simplicity and high level of versatility. At UC Davis our primary goal is to use the hybrid electric concept to create a vehicle with not only superior electric performance, but also with fuel economy surpassing the capabilities of conventional automobiles.

The powertrain consists of a single UNIQ SR180P 32kW, brushless, permanent magnet DC motor paired with a modified 20 hp (15 kW), 570 cc, four-cycle, air-cooled Briggs&Stratton overhead valve engine. We designed an innovative single shaft arrangement to mate both units to a Subaru 5-speed longitudinal transaxle driving the rear wheels. This results in a slender, long package well suited to our narrow track width and abundant length in the rear of the airfoil-shaped body.

Parallel vs Series Hybrid Powertrains

One of the critical decisions affecting the HEV design process is the decision to go with a series or hybrid configuration. The series hybrid uses the Alternative Power Unit (APU) as a generator, while a parallel hybrid directly transfer the torque of the APU to the drive wheels.

We decided on the parallel drivetrain in the earliest design stages. We carefully reviewed the advantages and disadvantages of each concept and compared them with our overall performance goals for the vehicle. The results are shown in Table 3.

Notice that on-board battery charging is listed as both an advantage and disadvantage. We do not consider this type of operation feasible, especially in California. The California Air Resources Board's reaction to hybrids is lukewarm at this time. They believe that owners of hybrids will not use them in their intended ZEV capacity, due to limited range and diminished performance.

Table 3

Series Hybrid

Advantages	1. Convenient on-board battery charging.
	2. APU can operate at best efficiency more easily.
	3. APU can cycle on/off easily.
	4. Transmission is not required for APU
	5. Lower peak power batteries can be used in vehicle.
	6. Ease of powertrain integration.
Disadvantages	1. Convenient on-board battery charging.
	2. Electric motor is the only powerplant contributing to vehicle acceleration.
	3. Fuel economy is not maximized because drive efficiency is poor.

A pure electric vehicle presents the owner with no other option but to religiously maintain and charge the battery pack at home. Our main publicity efforts will be targeted at changing this perception. Therefore, we try to avoid on-board charging the battery with the IC engine, except in unusual circumstances.

To illustrate a benefit of this point, consider the 7.1 kWh battery pack in the UC Davis HEV. Tests show that a full charge requires about 9 kWh of DC power from a charging system over six hours. The Solectria charger we use is over 93% efficient, requiring a total of 9.6 kWh of line voltage. Using the nighttime charging rate of $0.08/kWh (Sacramento Municipal Utilities District, off-peak rate), a full charge costs approximately $0.77. Charging with an APU at even 90% total efficiency would consume at least 1.12 gallons of gasoline, even at minimum specific fuel consumption, which is not feasible at a six hour charging rate. Thus, the same charge, done with the APU, is likely to cost at least $1.40 ($1.25 per gallon).

Setting aside the APU charging controversy, series hybrids do have many advantages, which why they are so common among automaker concept cars. The APU can easily operate at preferred speeds and loads, lowering average specific fuel consumption and avoiding some emissions problems. The powertrain is mechanically simpler and typically lighter than a parallel powertrain, if the same APU and electric motor are used.

A series powertrain requires no mechanical interaction between the electric motor and APU, so there is no need to address the problems of integrating the output shafts of the two. An effective series design will typically be lighter, simpler, and, therefore, more reliable than the equivalent parallel powertrain. The series drive vehicle allows for more flexibility in selecting batteries. The designers can specify batteries with a lower peak power rating than the peak power consumption of the electric motor. Nickel-Cadmium batteries have an inverse relation between peak power capacity and total energy storage. If the designers choose to forego maximum acceleration in ZEV mode, they can use an electric motor whose capacity is only attained with both the APU and the battery pack together supplying full power. This can enable the vehicle to extract slightly more range with the same battery weight.

The new Volvo ECC diesel hybrid concept car is a good example of series design. The NiCad batteries have an extremely high energy density of 48 Whr/kg, but a peak output of only about 114 W/kg for a 770 lb pack [10]. The diesel turbine genset supplies the additional 30 kW for acceleration. In comparison, the NiCads used in the UC Davis hybrid rated at a nominal energy density of 30 Whr/kg, but the usable peak power output is over 200 W/kg. The Volvo's acceleration times are markedly different as well: 0-62 mph in HEV mode takes 13 seconds, as opposed to 23 seconds in ZEV mode. The Volvo design, while a proven ULEV vehicle, offers very little hope of increasing the ZEV credits awarded to HEV's in California because its hybrid scheme does not allow full acceleration in ZEV mode.

It became apparent at the early stages of our calculations and simulations that limitations in the series design concept were steering us away from our goal of maximum fuel economy. The parallel hybrid, although more challenging in its execution, offers certain advantages that fit our requirements. Table 3.1 lists these tradeoffs.

The parallel hybrid concept, when properly executed, maximizes the versatility of the vehicle in question. The vehicle in hybrid mode drives like a conventional car, with the throttle pedal directly controlling the IC engine. Also, the APU can add its power to the electric motor during full HEV acceleration. Our vehicle must appeal to the consumer as well as our own California Air Resources Board. Therefore, we chose a parallel hybrid powertrain design as the best method to appeal to both groups. Our research and development efforts were aimed at solving the integration problems between the APU and electric motor. This should create a smooth power flow through the transmission, improving driveability. Careful engine management and powertrain control can reduce emissions problems, while the vehicle's extremely high fuel economy lowers the total grams/mile produced.

Table 3.1

Parallel Hybrid

Advantages	1. APU power transmitted directly to drive wheels.
	2. Improved acceleration.
	3. Range and driveability similar to a conventional car.
	4. APU recharging possible, *when needed*.
Disadvantages	1. Increased drivetrain complexity.
	2. Multi-speed transmission needed for IC engine to be effective.
	3. Emissions control of APU is more difficult, especially for NO_x.

Parallel Hybrid Electric Powertrain Development

Our efforts in powertrain development resulted in a lightweight, innovative parallel hybrid design capable of driving our energy efficient hybrid vehicle to its EV range goal of 60/60 and APU fuel economy of 100 mpg at 60 mph. The entire package, including IC engine, transmission, and electric motor/controller weighs only 380 lbs. and has a peak output of 134 ft-lbs of torque and 100 hp at the input shaft of the transaxle. The modified Briggs & Stratton engine is mated to the input shaft with an electromagnetic clutch adapted from a Nissan Continuously Variable Transmission (CVT). The Uniq Electric motor mounts outboard and transfers power through a 1.55:1 Gates™ cog drive belt. The transmission is a Subaru 5-speed longitudinal transaxle. The entire assembly mounts behind the passenger compartment driving the rear wheels.

Electric Motor Selection

The electric powertrain consists of the electric motor(s), any necessary converters or controllers, the battery pack, the battery charger, and any necessary safety equipment. We made an early decision to commit our research and development efforts to other sections of the vehicle. There are many excellent electric drive components available, prompting the decision to use available prototype or mass produced components. We attempted to find the best equipment currently available. This search resulted in a drivetrain that, while too expensive to be commercially viable at this time, represents what we feel 1998 model year electric vehicles will feature as standard equipment.

The electric motor used is Unique Mobility's UNIQ SR180P. This DC brushless motor main strengths are a sophisticated controller and extremely high peak specific torque. Due to the low power requirements of our vehicle in steady-state cruising (less than 5 kW at 60 mph), peak torque for acceleration is far more important than steady-state power. The motor uses the UNIQ CR20-300 controller. This 200 Volt solid-state controller has a peak output of 300 Amps. Thus, the system has an intermittent capability of 60 kW. We selected the UNIQ over its nearest competitor, the Solectria DC brushless. Solectria and Unique were chosen for their race experience and reputation for custom building dependable, high specific power electric DC motors. We summarized the performance of the UNIQ SR180P/CR20-300 system against two Solectria BRLS11/BRLS240H systems in Table 4. The Solectria motor/controller is much smaller, so a system of two are used for comparison.

Although the UNIQ motor is more than 25% more expensive, its peak torque characteristics make it a very attractive component for our system. It was important to remember when designing our electric drivetrain that this vehicle uses a relatively small IC engine, so acceleration depends primarily upon the electric motor. The CR20-300 also adds an integral, speed-sensing regeneration function. A two-motor system like the Solectria adds the capability of running only one motor during steady speed cruising. This can improve the efficiency of the motor/controller, but adds packaging and mounting complexity.

Main Battery Pack

Battery selection rates as possibly the toughest challenge facing any electric vehicle design. The viability of any electric or electric hybrid vehicles depends upon the emergence of a high power, high energy density, long life, affordable battery. Hybrid vehicles in particular require a higher power density due to the typically reduced size of their battery packs. Our own stringent weight goals also mandated a high energy density. Extensive testing by Argonne National Laboratory eased the task of selecting the appropriate battery. Some very promising prototypes exist, including Sodium/Sulfur, Zinc/Bromine, and the Ovonic Nickel/Metal Hydride battery. Unfortunately, none of these batteries are even commercially available, let alone affordable. This leaves us with a simple decision between Lead-Acid and Nickel/Cadmium batteries.

Flooded Nickel/Cadmium batteries and sealed Lead-Acid cells have markedly different characteristics. Flooded NiCads have extremely high power density, but are costly and require periodic maintenance (watering). Sealed Lead-Acid are less expensive, maintenance free, but have problems with power density and lifespan. The ANL tests reveal this relationship between SAFT STM5-200 Nickel/Cadmium batteries and Sonnenschein 6V160 sealed Lead-Acid [11].

Table 4

	UNIQ	Solectria (2)
Motor Weight	52 lbs	64 lbs
Controller Wt.	48 lbs	30 lbs
Cont. Power	32 kW	16.4 kW
Cont. Torque	34 ft-lb at 6,600 rpm	22.1 ft-lbs at 6,000 rpm
Intermittent Stall Torque	66.7 ft-lbs	47.2 ft-lbs
Peak Power	~ 60 kW	29.8 kW
Max Speed	7,500 rpm	7,000 rpm
Peak Efficiency	93%	94%

Table 5

	Lead-Acid	NiCad
Energy Density	36 Whr/kg	55 Whr/kg
Power Density	91 W/kg	175 W/kg
Testing Lifespan	370 cycles	>900 cycles

Due to the exceptionally good lifespan and high specific power, we chose Nickel/Cadmium cells from SAFT. These small, 40 Ahr cells have a reduced real-world energy density of 30 Whr/kg. This reflects the difficulty of combining 140 individual cells in series. Following the Argonne tests, which are based a driving cycle, the NiCad batteries in our HEV should last for at least 50,000 miles. The tolerance of the NiCad cell for high current draw and repeated deep discharging fits our need. In fact SAFT recommends and endorses full 100% DOD runs without worrying about shortening their lifespan.

The battery charger is a 2000 watt Solectria BC2000. The high voltage DC charger is over 93% efficient, and regulated by a 0-5 volt signal. At a charging rate of 10A, our batteries will reach full voltage in only four hours. Adding a two hour trickle charge at 4A enables a commuter to completely recharge the pack during working hours.

Alternative Power Unit

Our APU is based on a Briggs & Stratton overhead valve Vanguard four-cycle engine. This powerplant is subsequently modified to include overhead camshafts, new combustion chamber, electronic fuel injection and ignition, and modern emissions control.

We redesigned of the "top end" of the motor in an effort to reduce fuel consumption and extend the usable RPM range of the motor from 3600 rpm to about 5000 rpm. Our electric motor is reduced by 1.55:1 to the input shaft, so motor and the APU operated in approximately the same range. The smaller, high swirl, two-valve combustion chamber is based on a well-tested, successful Supermileage vehicle design. This design, in a 100 cc Briggs&Stratton engine, achieved minimum specific fuel consumption of 0.4, and real-world fuel economy of 3,313 mpg in our one-person vehicle, Side FX.

Preliminary testing of our modified motor is depicted in Figure 4, with some allowances for the inaccuracies introduced by our 175 hp dynamometer.

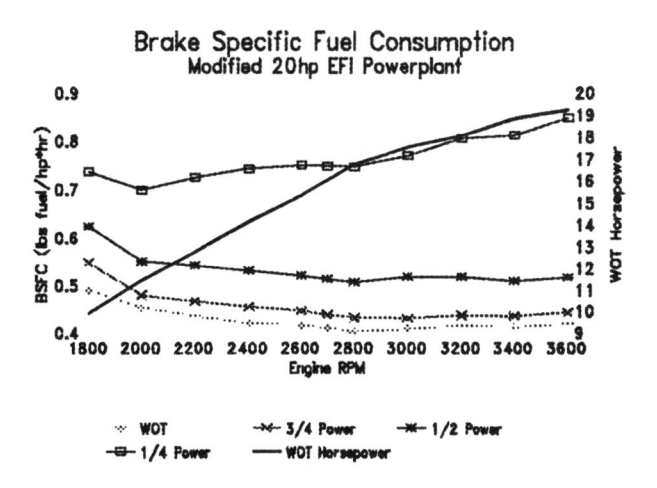

The stock 18 hp Briggs & Stratton Vanguard engine that we based our design upon features a relatively flat torque curve. This results in an engine with a broad efficiency range. Considering the engine's power curve, it is clear that the engine will operate at a very low specific fuel consumption due to its size. With the electric motor providing up to 100 ft-lbs of torque at the input shaft, we can concentrate on high efficiency, rather than high specific power for our APU design. This is really the only way to meet the lofty 100 mpg fuel economy goal.

The APU will also provide enough excess power to drive our lightweight vehicle in most circumstances. In fact, the vehicle can almost meet the Hot 505 cycle in APU-Only Mode as seen in Figure 5, below. The motor can meet all except the sharpest acceleration peaks.

Engine Management and Emissions

With the low-power IC engine, careful fuel management is critical. Our student-designed fuel injection system incorporates a hot-wire mass airflow sensor in conjunction with one heated oxygen sensor at each cylinder's exhaust port. Our own fuel-injection programming operates the multi-port, Bosch fuel injection system through an eight bit, C-programmable microcontroller. This custom building of the system allows us to tailor its operation to our specific engine management strategies.

Fig. 5
EPA 505 Driving Cycle
I/C Engine Only

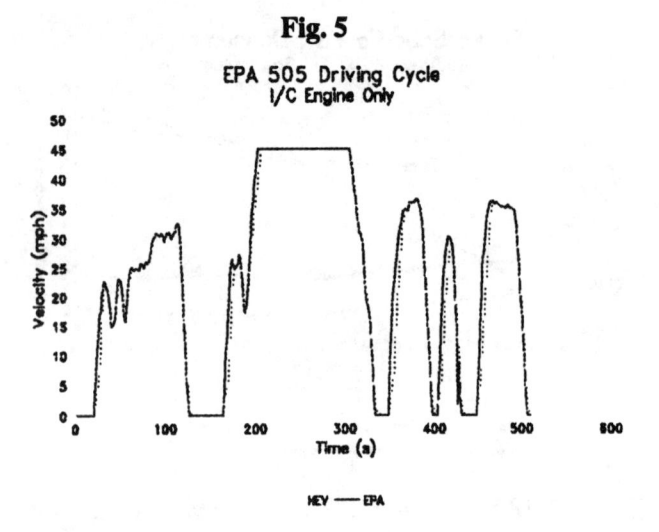

Exhaust emissions are regulated by a combination of active Exhaust Gas Recirculation (EGR) valves for each intake port and a custom three-way exhaust catalyst. With the small engine average cylinder pressures will be higher, forcing utmost care to be taken to control NO_x. Therefore, each cylinder has an independent HEGO sensor and pulse-width modulated egr solenoid. The catalyst is constructed to maintain a high average temperature with the lower displacement of our engine.

A standard evaporative vacuum cannister will reduce emissions from the fuel tank and the intake manifold. Low levels of fuel evaporation are expected as the engine management controller will usually choose not to run the APU below 1000 rpm.

Powertrain Control Strategy

Our parallel configuration allows us tremendous flexibility in exploring the various driving strategies and emissions controls ideas. For example, our APU will run off a high-speed start from the electric drive motor. Our electromagnetic clutch can engage smoothly to quickly start the IC engine. The fuel management system, sensing engine speed above 1500 rpm, can delay fuel injection turn-on until the IC motor is up to speed, avoiding startup emissions. Thus, we can quickly bring the IC engine up to the drivetrain speed, extracting power much more quickly than with a conventional start.

We plan two primary driving configurations, electric-only and HEV mode with the APU providing most of the power. In ZEV mode, the electric motor drives the car until the powertrain microcontroller's battery monitoring routine detects a 80% DOD. HEV mode can be immediately and passively triggered by the controller. Then the APU will high-speed start off the electric motor. In HEV mode, the APU provides almost all the power. The electric motor begins to provide power only for acceleration, not steady-state. The engine controller will always reserve 20% of the battery capacity for this type of acceleration. This allows the car to attain almost unlimited range, while maintaining high acceleration capability in each mode.

Although the Hot 505 driving cycle will be run in APU-only mode, Figure 6 offers an initial insight on the interaction of the two powerplants.

Figure 6
EPA 505 Driving Cycle
Hybrid Mode - Clean Start

The electric motor's high torque peaks (remember the 1.55:1 reduction to the drivetrain) cover most of the acceleration while the APU supplies steady-state power. This illustrates the advantage the HEV has in city driving. Stop-and-go acceleration and regeneration are supplied by the electric motor/controller. Longer trip highway speeds will bias the power supply towards the APU, as evidenced by the extended time at 45 mph on the Hot 505.

Conclusion

The intensive twelve month design and development process at UC Davis has yielded a versatile, promising hybrid electric vehicle. The low curb weight of 1800 lbs, aerodynamic efficiency, and parallel drivetrain have enabled this vehicle to reach its performance goals. [12] This can shed new light on the entire hybrid electric vehicle concept. After meeting its goals and performing in the HEV Challenge, we may have a vehicle capable of driving from our state capital in Sacramento to Los Angeles at 100 mpg, and then switching to ZEV mode and crossing the 60 mile LA. Basin on pure electric power.

References

[1] Abbott and von Doenhoff, *Theory of Wing Sections*, Dover Publicatins Inc.: New York, 1959, 119-122.

[2] Shevell, Richard, *Fundamentals of Flight*, Prentice Hall: New Jersey, 1989, 168-169.

[3] Waters, D.M., "Thickness and Camber Effects on Bodies in Ground Proximity", *Advances in Road Vehicle Aerodynamics*, 1973.

[4] Hoerner, S.F., *Fluid-Dynamic Drag*, pub. by author, 1965, 12.1-12.8.

[5] Dorgham, M.A., *Impact of Aerodynamics on Vehicle Design*, SP3 1983, 17-43.

[6] Hibbs, Bart, *GM Sunraycer Case History*, Lecture 2-2, SAE 1988, 18-43.

[7] Ashley, Steven, "GM's Ultralight is Racing Toward Greater Fuel Efficiency", *Mechanical Engineering*, May 1992, 64-67.

[8] Hartemann, Francois and Bernard Favre, "Human Factors for Display and Control", SAE 901149.

[9] Bosch Automotive Handbook, Ed. 2, 1986, 257

[10] Simanaitis, Dennis, "Volvo ECC", *Road & Track*, June 1993, 120-124

[11] DeLuca, WH, "Results of Advanced Battery Technology Evaluations for Electric Vehicle Applications", SAE 921572

[12] Winkelman, JR and Frank, AA, "Computer Simulation of the University of Wisconsin Hybrid Electric Vehicle Concept", SAE 730511

940510

Development of a 24 kW Gas Turbine-Driven Generator Set for Hybrid Vehicles

Robin Mackay
NoMac Energy Systems, Inc.

ABSTRACT

A 24 kW, 30% efficient gas turbine-driven generator set is being developed for hybrid electric vehicles. With the generator mounted on the same shaft as the compressor and the turbine, the rotor group is the only moving part other than the fuel pump. There are no engine-driven accessories. There is no oil as the rotor is mounted on air bearings. There is no water or antifreeze as the generator set is air cooled. Orientation can be vertical or horizontal.

Emission levels on unleaded gasoline, methanol, M85, ethanol or natural gas are much less than ULEV standards using a lean premix combustor. They are one or two orders of magnitude less than ULEV standards using a catalytic combustor. This paper addresses some of the design and application factors considered in developing the hardware.

INTRODUCTION

The use of electric vehicles is being encouraged by federal and state governments throughout the world and in many cases mandated. The primary objective is to reduce automotive emissions and therefore improve air quality. Yet, to achieve widespread commercial and consumer acceptance, range must be substantially improved.

Inadequate range is the most serious problem faced by electric vehicles today. Range requires energy. Batteries are the most common form of energy storage in electric vehicles. But batteries are heavy, so heavy that it is impractical to have a battery large enough to provide range similar to that of a conventionally fueled vehicle. In addition, the weight of the vehicle itself must be higher as a heavier chassis and larger tires are necessary to support the weight of the battery. Even short range electric vehicles must devote 30% or more of the vehicle weight to the battery.

The final result is a vehicle with limited range and reduced performance.

When air conditioning or electric heating is used, the range is further reduced. Depending on temperature, load and driving cycle, the reduction can be as much as 40%. In the case of a fully loaded bus operating in heavy traffic on a hot, humid day, far more energy will be used by the air conditioning than by the propulsion motor resulting in very short range.

The obvious solution to the range problem is the addition of a fuel-fired auxiliary power unit (APU). However, powering the APU with a conventional reciprocating engine reintroduces the emission problem when clean air is the primary rationale for electric vehicles. It also reintroduces the requirement for scheduled maintenance such as oil changes when minimizing service needs is an important secondary reason to use electric vehicles. Finally, a reciprocating engine-powered APU increases the weight of the vehicle significantly and uses up available cargo or passenger space.

This paper discusses the development of a lightweight, compact, 24 kW gas turbine-driven generator set. When equipped with a conventional combustor, the generator set weighs 41 kg (90 lb). It measures 66 cm (26 in) long by 41 cm (16 in) in diameter. The catalytic combustor increases the length by 25 cm (10 in). The NOx and CO emissions are less than 20% of those projected for electric utility plants in the Los Angeles area producing equivalent amounts of electricity in the year 2010.

DESIGN PHILOSOPHY

The original design objective was to create a low cost, low maintenance generator set for cogeneration and other continuous duty applications. Thus the design had to be extremely conservative using a minimum number of parts and those parts had to be inexpensive. The unit would have to run for several years without scheduled maintenance.

Three major changes were made to adapt the generator set concept to hybrid vehicle applications. A smaller recuperator is used to save weight which reduces fuel efficiency from 35% to 30%. An unleaded gasoline combustor is being developed as gasoline is clearly the most readily available fuel. Finally, different power conditioning electronics are needed as the electric output to the battery is direct current rather than alternating.

The compressor and turbine wheels are quite conventional. However, the rest of the generator set varies from common practice. There are no gears and no engine driven accessories. There is no lubrication system. There is no separate starter motor. There is no fire in the combustor. The key engineering advancements can be summarized as follows:

1. The combustor uses a flameless catalytic reaction and a high air/fuel ratio of 134:1. This reduces the NOx, CO and HC emissions to a small fraction of the California ULEV requirements.

2. The generator turns at 96,000 rpm and is mounted on the same shaft as the compressor and turbine. This eliminates the need for a reduction gear box. It also reduces the size and weight of the generator as the output is a direct function of rpm.

3. The rotor group is the only continuously moving part other than the fuel pump. It is mounted on self pumping air bearings thus eliminating the need for a lubrication system with its attendant service requirements.

4. The recuperator is wrapped around the generator set reducing the overall generator set dimensions to 91 cm (36 in) long by 41 cm (16 in) in diameter while holding the weight to 41 kg (90 lb).

RATING

Most vehicles sold in the United States have engines rated between 75 kW and 150 kW. It would be quite feasible to develop a gas turbine-driven generator set with a rating in this range and use it in conjunction with an electric motor drive. The vehicle would then be conventional with two exceptions. The power plant would be a gas turbine rather than a reciprocating engine and the power would be transmitted electrically to the wheels rather than mechanically.

Two problems arise immediately. First, there would be no regenerative braking as there would be no propulsion energy storage device. Second, the generator set would operate at roughly 10% power when cruising which is an inefficient operating point for a gas turbine.

A rating of 24 kW, continuous, was selected which would be adequate for both passenger car and small bus (22 passenger) applications. This power allows a 1300 kg (2860 lb) vehicle with a low drag coefficient to climb a 6% grade at 100 km/h (62 m/h) on generator set power alone. Energy storage in the form of a small battery or a flywheel will be used in conjunction with the generator set for regenerative braking and to provide power for rapid acceleration. The key is that the energy storage device must be designed for high power rather than high energy.

24 kW is also adequate for 22 passenger buses operated in the city in stop-and-go traffic with air conditioning. For example, Santa Barbara, California has had at least eight electric buses in service for several years. Although they are smaller than standard buses, they are 22 feet long, have a GVW of 15,000 lb and carry 22 seated and 7 standing passengers plus the driver. The motor is rated at 30 kW continuous or 45 kW intermittent and the top speed is 64 km/h (40 m/h). The batteries are rated at 70.2 kWh and run the vehicle for over eight hours per day without recharging for an average load of less than 9 kW. The buses are not air conditioned but if 10 kW is added for electrically-driven air conditioning, a 24 kW generator set is still more than adequate.

The 24 kW generator set is ideal for those buses currently run on battery power alone. The proposed arrangement uses energy storage (generally batteries) for acceleration and hill climbing. When stopped, the generator set keeps running and charges the battery. During deceleration, energy is recovered which typically reduces the energy consumed during a day's run by about one third. Larger buses or buses operated at freeway speeds can use two, three or four units as the individual generator sets weigh only 41 kg (90 lb) each. Paralleling and load sharing are easy when the load is direct current.

The rating point is based on an ambient temperature of 35°C. (95°F.) at sea level. Although most gas turbines are rated at 15°C. (59°F.), it is obvious that the generator set will have to operate during summer months and/or southern climates for extended periods at higher temperatures. Note that the 35°C. (95°F.) refers to the air entering the generator set, not the compressor. The air passes through the generator to cool it before entering the compressor. Thus the air entering the compressor is warmer than ambient.

ROTATING ASSEMBLY

The generator rotor, compressor wheel and turbine wheel all turn at 96,000 rpm at full load. They are assembled together on a common shaft without gears and therefore turn at the same speed. The rotor assembly is shown in Figure 1.

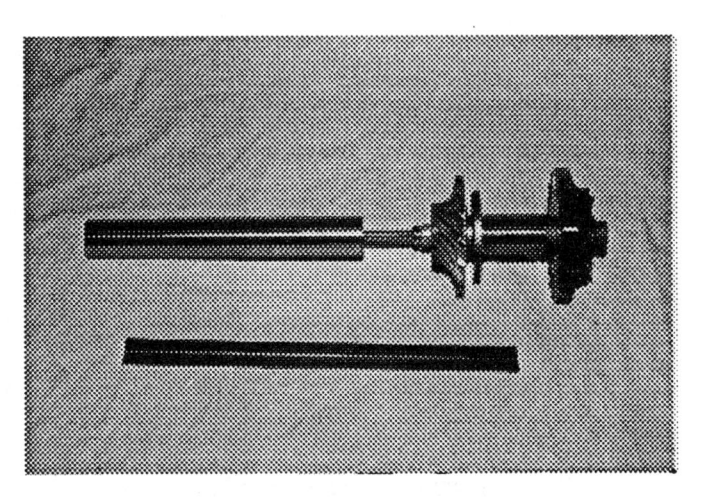

Figure 1
Rotor Assembly with 12 Inch Scale

The generator is mounted outboard of the compressor so that all of the air passes through it, cooling the generator before entering the compressor. As the generator is 96% efficient and rated at 24 kW, slightly less than 1.0 kW is dissipated as heat. With a mass flow of 15 kg/min (33 lb/min), the temperature rise through the generator is 4°C. (7°F.).

Although the increase in temperature reduces the power of the gas turbine by 400 watts, the loss is less than would occur with either a separate cooling fan or a liquid cooling loop. In addition, the elimination of cooling fans and pumps improves the reliability, while reducing the maintenance, volume, weight and cost.

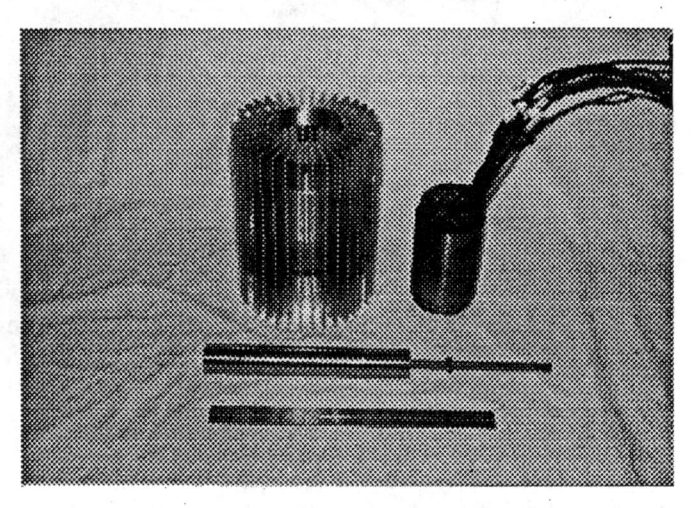

Figure 2
Rotor, Stator and Generator Housing with 12 Inch Scale

Mounting the generator outboard of the compressor solves another thermal problems. If the generator were mounted between the compressor and turbine wheels, some of the heat from the turbine wheel would be radiated and/or conducted into the generator rotor. Some form of auxiliary cooling would be necessary with accompanying power and efficiency losses to say nothing of increased complexity. The problem of cooling a generator mounted next to a turbine wheel is even more difficult after a normal shutdown due to heat soakback. After a hot shutdown, it is extremely difficult.

The generator is a permanent magnet design. The rotor, stator and housing are shown in Figure 2. Although neodymium boron iron was evaluated for use as the magnet material, it was decided to use samarium cobalt because of the ability to handle higher temperatures.

The compressor and turbine wheels (Figure 3) are similar to those used in turbochargers but higher in aerodynamic efficiency. The centrifugal compressor has a single stage with an efficiency of 79.6% and a pressure ratio of 3:1. The radial inflow turbine has a single stage and an efficiency of 85.0%. The turbine inlet temperature is 816°C. (1500°F.)

Figure 3
Compressor and Turbine Wheels with 12 Inch scale

The rotor group is mounted on three sets of radial air bearings with an additional pair of air bearings mounted on either side of a rotating flange to handle thrust. The bearings are self pumping and require no source of pressurized air. They locate the rotor group in a very positive manner so that the compressor tip clearance can be minimized, improving the compressor efficiency.

The air bearings offer several advantages. They completely eliminate the need for a lubrication system and therefore the need for oil changes and other lubrication service requirements. Cost is reduced because the air bearings are far less expensive than the combination of conventional bearings, oil pump, oil sump and oil cooler. Size and weight are also reduced because of the elimination of these parts. Finally, the generator set can be mounted horizontally or vertically.

RECUPERATOR

To achieve a projected efficiency of 30% from fuel-in to electricity out while still maintaining a low turbine inlet temperature, a recuperated or regenerated cycle is needed. Otherwise, the efficiency would be an unacceptably low 14%.

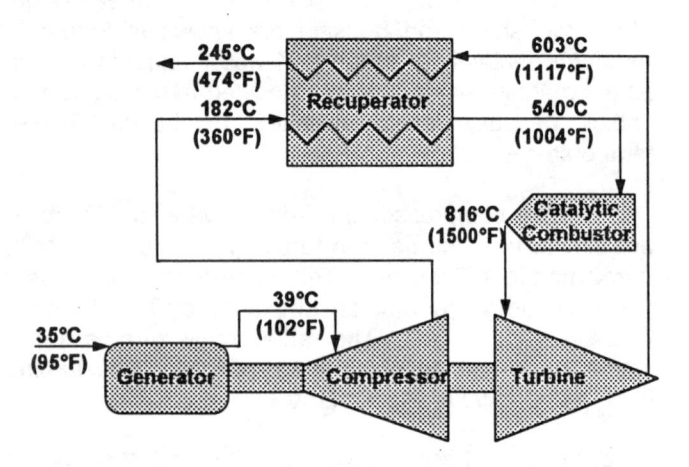

Figure 4
Flow Diagram with Temperatures (Calculated)

Recuperators and regenerators are nothing more than heat exchangers which utilize the heat of the exhaust to preheat the air going to the combustor. Thus only enough fuel is needed to raise the temperature of the combustion air to 816°C. (1500°F.) from 540°C. (1004°F.) instead of from 182°C (360°F.). This is shown in Figure 4. Fuel consumption is reduced by 60% raising the thermal efficiency by a factor of 2.5 as compared with an unrecuperated cycle. This cycle works best where the pressure ratio is low because a low pressure ratio results in a low compressor discharge temperature and a high turbine discharge temperature. This permits more heat to be transferred.

Most automotive gas turbines use regenerators which are heat wheels that rotate through the turbine exhaust and the combustor inlet air. They offer the advantage of compact design and high effectiveness. However they suffer from rotating seal leaks which impact performance negatively and from seal wear which shortens life and increases maintenance.

Most stationary gas turbines that use this cycle have plate fin or tubular recuperators. These offer long life and do not have a seal problem but tend to be very heavy and bulky. The recuperator selected for the gas turbine discussed in this paper is an all prime surface, counterflow recuperator with an effectiveness of 85%. It is wrapped around the generator set. Both the exhaust gas and the combustor inlet air flow circumferentially through the recuperator. The result is a recuperator that is substantially lighter and more compact than either conventional recuperators or regenerators. As nothing rotates, there is no dynamic seal or leak problem. In addition, there is virtually no manifolding, so the pressure drops are low.

Note that the recuperator is a major factor in selecting the turbine inlet temperature. Although increasing the turbine <u>inlet</u> temperature results in better efficiency and higher specific power, it also results in a higher turbine <u>discharge</u> temperature. As the turbine discharges into the hot side of the recuperator, increasing this temperature can force the use of far more expensive materials in the recuperator to prevent premature failure. If the end objective is to mass produce generator sets for hybrid electric automobiles, cost is one of the most critical issues.

COMBUSTOR AND EMISSIONS

Spark ignition engines produce a substantial amount of NOx which is then cleaned up as much as possible in a catalytic converter. Gas turbines produce far less NOx, especially if they use a lean premix burner or if they have water or steam injection. However, the amount of NOx is still significant because the peak temperatures in the primary zone of a conventional combustor can easily exceed 2000°C. (3632°F.).

Figure 5
Catalytic Combustor

The gas turbine being developed uses the catalytic combustor shown in Figure 5. It works on an entirely different principle from a conventional combustor. Fuel and air will react in the presence of a catalyst to produce heat without a flame. The temperature in the catalytic combustor does not exceed 850°C. (1562°F.) which is well below the threshold where thermal NOx is formed. The catalyst is used to produce heat with virtually no NOx formation rather than to clean up the NOx that was formed in engines or gas turbines with conventional combustion.

Starting the catalytic combustor is relatively straight forward. The leading edge of the catalyst bed has a resistance heating element in it. This is used to heat the catalyst during starting, after which it is turned off. The reaction is then self sustaining.

Preliminary testing on the catalytic combustor has shown that virtually no NOx is produced. The first tests were run with propane as a fuel. They showed a range of from 0.09 to 0.19 ppmv, uncorrected, or 0.54 to 1.14 ppmv, corrected to 15% oxygen. The maximum reading correlates to 0.026 g/kWh (0.019 g/hp/hr). Based on 130 Wh/mi, which is the energy consumption quoted by Road & Track magazine (reference 5) for AC Propulsion's Honda CRX driven at 97 km/h (60 m/h), the NOx emissions would be 0.0034g/mi. Although California's ULEV standard cannot be used as a direct comparison because it is based on a specified driving cycle and the CRX emissions are based on steady state driving, the ULEV standard is 0.200 g/mi. Note also that it is extremely difficult to measure NOx levels that are that low. The readings noted are within the noise level of most of the instrumentation used to measure NOx.

The catalytic combustor should also do well on CO emissions where it is helped by the 134:1 air/fuel ratio. However the CO emissions on the first runs were higher than expected because of the test setup used. These tests are being rerun and the results will be presented when the paper is given.

Combustion efficiency was in excess of 99%. As long as the combustor outlet temperature was in excess of 755°C. (1391°F.), the hydrocarbon levels were too low to be read on the instrumentation.

Testing is being continued on the levels of all three pollutants using various fuels. More detailed results will be available when the paper is given.

The gas turbine will also be tested with a conventional lean-premix, multi-fueled, annular combustor. It will have low emissions but not as low as those from the catalytic combustor. However, it will be less expensive.

The generator set will be produced with the option of a catalytic or a conventional combustor. Both versions will be substantially better than California ULEV requirements. However, only the catalytic combustor version will have emissions lower than 20% of those projected for electric utilities in the Los Angeles Air Basin in the year 2010 (reference 3). But it still will not be classed as "zero emission". Only pure electric vehicles can achieve this status.

Electric vehicles can achieve it only because the emissions from the utility are ignored by the regulations. Unfortunately, at the present time, there is no category between ULEV and "zero emission". If ULEV is met and "zero emission" is not, there is no legal incentive to lower emissions. There are indications that these regulations may change and that total emissions will be considered so that there will be a greater reason to purchase a vehicle with the catalytic combustor.

CONTROLS

The output of the generator must match the needs of the battery. Typically, the battery will have a nominal voltage of perhaps 336 volts. However, this can drop by more than 100 volts during acceleration, especially when the battery is partially discharged. At the other extreme, in a regenerative braking mode, the battery voltage can be much higher than nominal.

The first challenge then is for the generator set to provide the appropriate power at the appropriate voltage, recognizing that with a permanent magnet generator there is no voltage control. In response to a signal from the vehicle, the generator set puts out the required power at a voltage that is always lower than the minimum battery voltage. This voltage is then electronically boosted to match the requirements of the battery.

Response time in terms of changing the power output is not critical. It is assumed that if the driver demands maximum acceleration, the battery will provide the power. The generator set can then increase its power output over a period of a few seconds without creating an operational problem.

The generator set is fully capable of running continuously at any load from 0 to 24 kW. 24 kW is produced at 96,000 rpm with a turbine inlet temperature of 816°C. (1500°F.) on a 35°C. (95°F.) day at sea level. If less power is needed, the turbine inlet temperature can be reduced. This is the way it is done in most gas turbines but the lower turbine inlet temperature impacts fuel efficiency drastically at part load.

A better way to reduce power is to lower the rotor speed. This reduces the mass flow and the pressure ratio. If the turbine inlet temperature is held constant, the reduction in fuel efficiency will not be nearly as great as it would be if the turbine inlet temperature were lowered.

Rotor speed is controlled in the generator set by simultaneously adjusting the fuel flow into the gas turbine and the power taken out of the generator. Generator output is controlled by varying the amount by which the output voltage is boosted.

TEST PROGRAM

The first three units are in various states of assembly, disassembly and test. One of these will be shipped to a major automobile manufacturer for bench testing followed by a test program in a hybrid electric vehicle. The other two units will be used internally for test, evaluation and further development.

Starting in late 1994, thirty units will be shipped for field test. Many of these will go into hybrid electric vehicles which will provide operational experience.

Buses are the preferred vehicles as they tend to be more expensive than automobiles or light trucks and can therefore justify the higher cost of prototype hardware. In addition, they have professional drivers, are usually supported by a crew of mechanics and generally return to a maintenance facility every day.

An additional group will go into stationary applications such as cogeneration or air conditioning drives. These applications will accumulate many hours at full load and many starts.

Installations will be selected based on the following criteria. Good support must be available locally and it must be easy to provide factory assistance. The application should be one with the potential of follow-on business. There should be an obvious need for the turbine.

SPECIFICATIONS AND PERFORMANCE

The generator set specifications are given in Table 1. Power is measured at the generator terminals. Efficiency is based on the lhv of the fuel versus electricity at the terminals.

power	24 kW, continuous
rating conditions	35°C. (95°F.), sea level
efficiency	30% @ 24 kW
rpm	96,000
pressure ratio	3:1
turbine inlet temp	816°C. (1500°F.)
exhaust gas temp	255°C. (491°F.)
compressor	single stage, centrifugal
copressor effic.	79.6%
turbine	single stage, radial inflow
turbine efficiency	85.0%
generator	permanent magnet
magnet material	samarium cobalt
generator effic.	96%
recuperator	circumferential, fixed boundary
rec. effectiveness	85%
fuel, gaseous	natural gas, propane, butane
fuel, liquid	gasoline, methanol, M85, ethanol
length	66 cm (26 in) (conv. combustor)
length	91 cm (36 in) (catalytic combustor)
diameter	41 cm (16 in)
weight	41 kg (90lb)
orientation	horizontal through vertical

Table 1
Generator Set Specifications

COST

The generator set has the potential for being very low cost. The individual components can be compared to those of a conventional turbocharged engine. The compressor and turbine wheels are similar to those in a turbocharger. The materials are the same. The turbine inlet temperature is actually lower. The aerodynamics are different but not more expensive. The air bearings are lower in cost than the combination of lubricated bearings, oil pumps and oil coolers.

The generator is comparable in size to the alternator on a conventional engine. It produces far more power but that is because it is turning roughly twenty times as fast. The catalytic combustor is similar to the catalytic converter on a conventional engine.

The gas turbine generator set still has to account for the recuperator and the power electronics. However, the cost of these components must be matched against the reciprocating engine's block, head, pistons, connecting rods, crank shaft, camshaft, valve train and radiator as well as accessories such as the water pump, fan, starter and, of course, the various computers.

The initial thirty units will not be inexpensive. However, starting in 1995, early production units should be fully competitive for bus applications. When full production is reached, these generator sets should be much less expensive than those powered by reciprocating engines and should be priced appropriately for installation in automobiles.

CONCLUSION

If "zero emission" vehicles are to have a major impact on air quality, they must be purchased in great numbers. In other words, the consumer and the commercial user must want them. Yet most electric vehicles that are proposed or are on the market today offer limited range, long battery recharge times and slow acceleration. Air conditioning and heating, if available, are often marginal. Both the capital cost of the vehicle and the cost of replacing batteries are excessive. It is a challenging task to sell these features to customers in the name of protecting the environment.

On the other hand, consider the concept of a hybrid electric vehicle equipped with a small gas turbine-powered APU using a catalytic combustor. It can offer performance and range that equals or exceeds that of a conventional vehicle. In reference 5, the fuel consumption of AC Propulsion's 1329 kg (2930 lb) Honda CRX driven at a steady 97 km/h (60 m/h) and equipped this way was calculated. The fuel consumption was between 3.59 L/100 km (65.5 mi/gal) and 3.14 L/100 km (74.9 mi/gal). Yet this vehicle in its all-electric mode demonstrated 0 to 97 km/h (0 to 60 m/h) acceleration in 7.8 seconds (reference 5).

The range and performance achieved this way make the vehicle highly desirable, especially as the comfort of full air conditioning and heating can be offered. Yet this combination would offer lower emissions than those of a pure electric vehicle powered by a battery that is recharged from the electric utility.

REFERENCES

(1) R. Mackay, J. Noe, "High Efficiency, Low Cost, Small Gas Turbines", ASME Cogen-Turbo V, Budapest, Hungary, 3-5 September, 1991

(2) R. Mackay, "Gas Turbine Generator Sets for Hybrid Vehicles", SAE International Congress & Exposition, Detroit, MI, 24-28 February 1992

(3) Bevilacqua Knight, Inc. "1992 Electric Vehicle Technology and Emissions Update", prepared for California Air Resources Board, Agreement No. A866-187, 30 April 1992

(4) R. Mackay, J. Arias, J. Rowlette, "Ultra Low Emission, Hybrid Vehicles using Micro Gas Turbines and Bipolar Batteries", 25th ISATA Silver Jubilee Conference, Florence, Italy, 1-5 June 1992

(5) K. Reynolds, "AC Propulsion CRX", Road & Track magazine, October 1992

(6) R. Mackay, "Hybrid Vehicle Gas Turbines", SAE International Congress & Exposition, Detroit, MI, 1-5 March 1993

940556

The NGV Challenge - Student Participation, Faculty Involvement, and Costs

Richard M. Lueptow
Northwestern Univ.

ABSTRACT

Automotive design projects offer an outstanding opportunity for students to obtain practical experience in engineering design. Over the last three years student groups have competed in one such project, The Natural Gas Vehicle Challenge, which involved converting a pickup truck to be fueled by natural gas. This paper reports a survey of faculty advisors involved with the project at the competing universities. Fifteen students were typically involved in the project at each university participating in the competition. Five or six were usually "key" to the project. Usually faculty advisors had a research interest in automotive engineering or alternate fuels, and they often incorporated the project into a design course. Although the funding level for such a design project varied substantially, the typical funding level for one year was about $25,000, most of which came from local sponsors. Faculty advisors often commented on the educational value of the project and their satisfaction in working closely with students.

INTRODUCTION

Automotive design projects in which students from different universities compete against one another in the design of a vehicle can be ideal design experiences for engineering students. The projects require that students approach design much as they would in industry. Students must address the realities of taking a design from the computer screen to the machine shop; they must cope with schedules and deadlines; they must meet design and safety specifications; they must implement a design that is robust and reliable; and they compete against other students who started with essentially the same resources. This paper addresses the student participation, faculty advising, and financial costs involved with competing in one such large-scale automotive design competition, The Natural Gas Vehicle (NGV) Challenge.

The Society of Automotive Engineers (SAE) is a leader in providing group design projects for engineering students. SAE sponsored competitions include the SunRayce, Formula SAE®, SAE Mini-Baja®, and alternate-fuel vehicle competitions such as The Methanol Marathon and The NGV Challenge. In all of these events SAE provides a format in which teams of students from various universities attempt to build or modify a vehicle while staying within a well-defined set of rules.

One recent competition for alternate-fuel vehicles was The Natural Gas Vehicle Challenge. The goals of The NGV Challenge included developing new technology for natural gas as an alternate fuel, encouraging interest in automotive engineering, and providing students with a realistic design experience. The competition was sponsored by General Motors Corporation, the U. S. Department of Energy, and Energy, Mines and Resources--Canada, along with SAE.

In 1990, proposals were sought from universities in Canada and the United States for converting a full-size GM pickup truck to be fueled by natural gas instead of gasoline. Twenty-four universities were chosen to receive a truck donated by General Motors. Over the next months, students worked to modify the fuel system, engine, engine control system, and emissions system to provide optimal performance using natural gas as a fuel. A year later, in June 1991, student teams met in Oklahoma to compete against one another in events rating exhaust emissions, fuel economy, design, and vehicle performance. The individual tests ranged from a full Federal Test Procedure emissions test to a road rally. Monetary prizes and trophies were awarded to the top teams. Most universities participated in the second year of the event in 1992 and the third year of the event in 1993. The technical results of the competition and the vehicle designs have been compiled (1, 2).

In this paper three aspects of participation in The Natural Gas Vehicle Challenge are addressed: 1) the degree of student participation and the coordination of the project with the engineering curriculum; 2) the degree of faculty involvement and how their participation relates to their other academic duties; and 3) the financial resources necessary to be involved in such a competition. A summary of comments by faculty advisors for the project is also included. While the competition for which advisors were surveyed was the 1992 NGV Challenge, the results have broader application to other group design competitions.

SURVEY PROCEDURE

Faculty advisors for all twenty-one universities competing in the 1992 NGV Challenge were surveyed about one month after the competition. Seventeen faculty advisors responded. The responses were anonymous. While anonymous surveys have the advantage of encouraging full disclosure, they have the disadvantage of not allowing follow-up to responses that seem unusual or need explanation. Several questions in the survey required faculty advisors to estimate the levels of participation or funding. Clearly, some estimates are more accurate than others, depending upon the records that were kept and the individual interpretation of the questions.

The responses to the survey were tabulated and provide the basis for this paper. Most questions on the survey required specific responses, although the survey included some open-ended questions. Only one survey was sent to each participating university, although several universities had multiple advisors.

STUDENT PARTICIPATION

The data for student participation in the project for the seventeen responding universities are shown in figure 1. An average of 15 students participated in the project at each university, although as few as four students and as many as 30 students took part at different universities. Of the students who participated, 49 percent worked on the project for credit in a regular course such as a capstone design course, but not including independent study courses. The distribution of students participating in the project and receiving regular course credit for their work is also shown in figure 1. At twelve universities The NGV Challenge project was incorporated into classroom instruction, typically for a design course or a special topics course. One university used the project in three courses, six universities used the project in two courses, and the remainder used the project for a single course. In 35 percent of the courses more than 75 percent of the students in the class worked on the project. In the case of two universities every student who worked on the project received course credit. At five universities the project was not related to regular coursework.

Students also received credit for their work on the project as independent study at nine universities. About 10 percent of the students participating in the project worked for independent study credit as an alternative to or in addition to regular course credit.

The competition rules limited the participation of graduate students to 25 percent of the team from any given university. Only one school had that level of graduate student participation, while only undergraduates were involved in the project at eleven universities. The overall graduate student participation was 5 percent.

Although students actually taking part in the competition were required to be SAE members, not all students working on the project were members of SAE. Overall, 83 percent of the students working on the project belonged to SAE, but the percentage for individual universities ranged from a low of 33 percent at one university to 100 percent at nine universities.

Most of the students participating in the project, 89 percent, were enrolled in the mechanical engineering department, followed by electrical engineering students with 6 percent of the participants. Students registered in other engineering curricula made up the remainder. Only one university had participation of any non-engineering students. Perhaps more interesting is the level of participation within a mechanical engineering department. The number of students in all classes, freshman to senior, participating in the project is compared with the number of graduating mechanical engineering students in figure 2. The number of mechanical engineering students graduating annually from participating

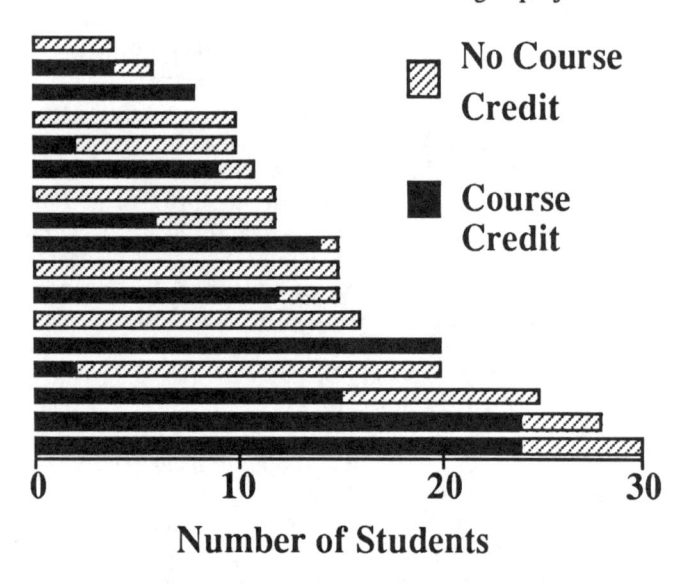

Figure 1: The total number of students participating in the project at each responding university is shown by the total length of the bar. The solid black portion of the bar indicates students who received regular course credit (not including independent study credit) for their participation in the project. Bars are ordered from smallest number of participating students to largest number.

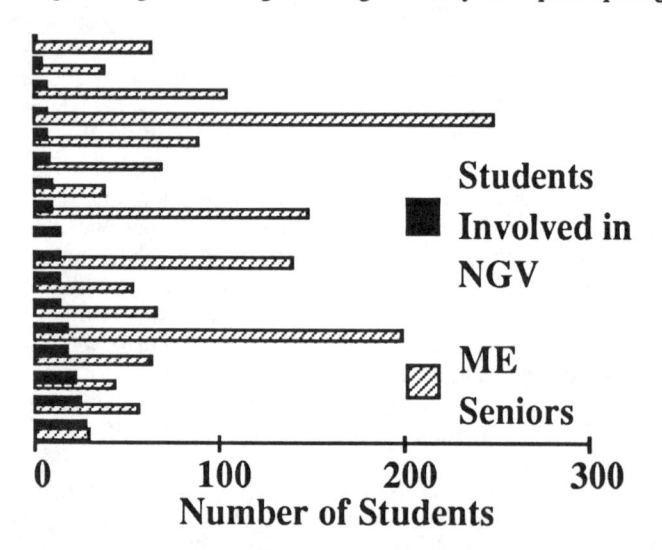

Figure 2: The total number of student participating in the project is indicated by the black bar. The number of graduating seniors at the corresponding university is shown by the cross-hatched bar. Bars are ordered from smallest number of participating students to largest number.

universities ranged from 32 to 250. But the number of mechanical engineering students participating was not related to the size of the department as measured by the number of graduating seniors. At some universities a very small number of students participated compared to the number of graduating seniors, and at other universities the number of students participating in the project was nearly equal to the number of graduating seniors.

The most telling statistic about the student participation is the number of "key" students who dedicated substantially more time to the project than other students participating. On the average five or six students were identified as "key" students, although the number ranged from 2 to 15. The number of key students was independent of the total number of students participating in the project (figure 3).

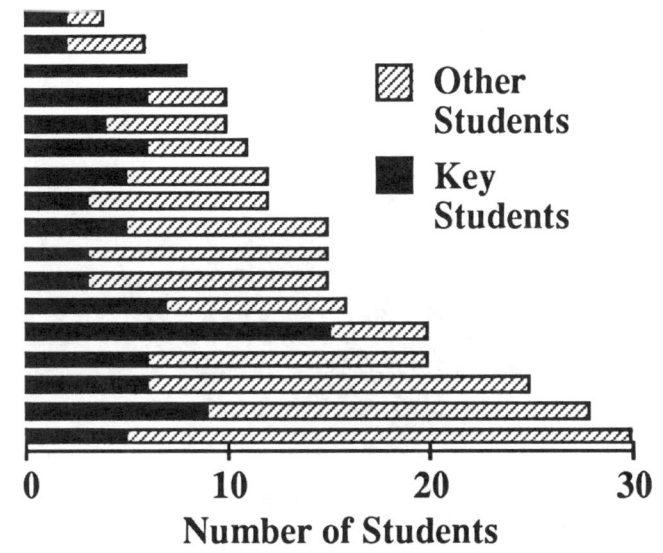

Figure 3: The number of "key" students participating in the project is indicated by the black portion of the bar, the total length of the bar representing the total number of students involved. Bars are ordered from smallest number of participating students to largest number.

These results indicate that the typical student group had about fifteen mechanical engineering students of whom five or six played a major role in the project. Most of students were SAE members, and about half received some sort of credit for their participation in the project. The project did not necessarily have broad interest within the mechanical engineering department. The ratio of participants to graduating seniors suggests that a very small percentage of the students in the mechanical engineering department were involved with the competition at some universities, while other universities had a substantial fraction of the mechanical engineering students involved.

FACULTY INVOLVEMENT

A requirement for participation in The Natural Gas Vehicle Challenge is the involvement of a faculty member in advising the team. Advising a team of undergraduates working on an open-ended design project requires substantial time and effort. This section of the survey explored the relationship of the advisor to the project.

At all universities participating in the survey, the project was centered in the mechanical engineering department. At thirteen of the seventeen responding universities a single advisor was involved with the project. Two of the remaining universities had two advisors, and two had three advisors. In all cases at least one of the faculty advisors was in the mechanical engineering department. In two of the programs with more than one advisor, the second advisor was from the electrical engineering department. All but three faculty advisors were tenured.

Most faculty advisors initiated their own involvement in the project because of their personal interest. In ten of seventeen programs the faculty advisor initiated the project either because of an interest in automotive engineering or engines or because of an interest in design. Five faculty advisors volunteered to advise the project when asked by students and two advisors volunteered when asked by the department chair or other faculty.

Faculty advisors undertook the responsibility for advising their university's entry in The NGV Challenge in addition to their own research programs. As a measure of the extent and interest of the faculty advisor's research program, the advisors were asked about the total number of graduate students they advised and the number of those graduate students who were pursuing research in automotive engineering, alternate fuels, or combustion (figure 4). (Because of the way that the survey question was asked, responses from universities with two or more faculty advisors were unclear and therefore omitted, leaving only thirteen responses shown in the figure 4). Most faculty advisors have active research programs, often in automotive engineering or a related field, so advising an automotive design projects fits in well with the focus in their research program. Several faculty advisors have large research programs and advise up to fifteen graduate students.

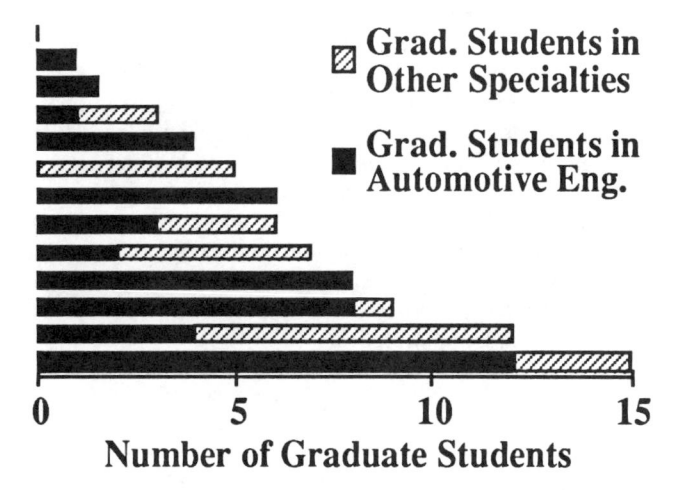

Figure 4: The research level of faculty advisors as indicated by the number of graduate students that they advise shown by the total length of the bar. The faculty advisor's interest in automotive engineering, alternate fuels, or combustion is indicated by the number of graduate students that they advise in those areas, shown as the black portion of the bar. Bars are ordered from smallest number of students advised to largest number.

The funded research programs of the faculty advisors ranged in size from zero to $750,000 annually. The NGV Challenge stimulated further interest in research related to natural gas vehicles. At seven of the seventeen responding universities participation in The NGV Challenge has brought about new research opportunities that are presently funded. Three of these universities plus seven others are currently seeking research funding as a result of their participation. In only three cases are the universities not seeking research funding for projects related to The NGV Challenge.

The role of the faculty advisor in the project varied substantially. At nine universities the faculty advisors wrote the original proposal for the university's entry in The Natural Gas Vehicle Challenge. In seven cases the students who eventually participated in the competition wrote the proposal, and in one case the students involved in another SAE competition wrote the proposal. At twelve universities the students organized and led the effort for the entry in the event with active involvement of the faculty advisor. In three cases the students led the entire effort with minimal involvement of the faculty advisor, and in two cases the advisor led the effort with active involvement of students. Students played a major role in the design of the vehicle. At five universities the students were responsible for the entire design; at ten universities the students were responsible for the design but had active input from the advisor; at two universities the advisor was primarily responsible for the design with active input from the students.

As mentioned earlier, at twelve universities work on The NGV Challenge project was incorporated into design or special topics courses. In nine of these twelve universities, the faculty advisor taught the course. But at only two of the seventeen universities did the faculty member have his teaching load reduced to compensate for time spent outside of the classroom advising students working on the project.

The typical model for faculty advising seems to be that the faculty member wrote the initial proposal but students followed through with the project organization and vehicle design. The faculty advisors typically took on the task of advising students because of their own interest in automotive engineering or alternate fuels as evidenced by their research programs. The faculty advisors were willing to advise the students working on the project in addition to their own research even with no teaching load reduction, although they sometimes used the project in their own design courses. Often, the faculty advisor used the project to enhance other research funding opportunities.

COSTS

One of the issues in carrying out a design project like The Natural Gas Vehicle Challenge is financing the design and development. Figure 5 shows the typical distribution of sources for the funds that universities used for vehicle development. The average cash funds received for the second year of the event was $24,750. Clearly the most important source of funding was local sponsors such as local gas utilities, gas pipeline companies, and state governments. The funds varied substantially from university to university (figure 6). The smallest funding level was $3,500, while one university had $56,000. Six universities had funding levels greater than $30,000 primarily as a result of local sponsorship. Four had funding levels under $10,000. Some of the inequity came about because Canadian teams received larger amounts from their federal sponsors than United States teams were given. The university contribution to the effort ranged from zero to $10,000, with an average of $3,100. Local sponsorship ranged from zero to $40,000. Clearly some universities had better connections with local industry and government than other institutions. Some teams also used their winnings of up to $5,000 from the previous year's event to fund their vehicle development.

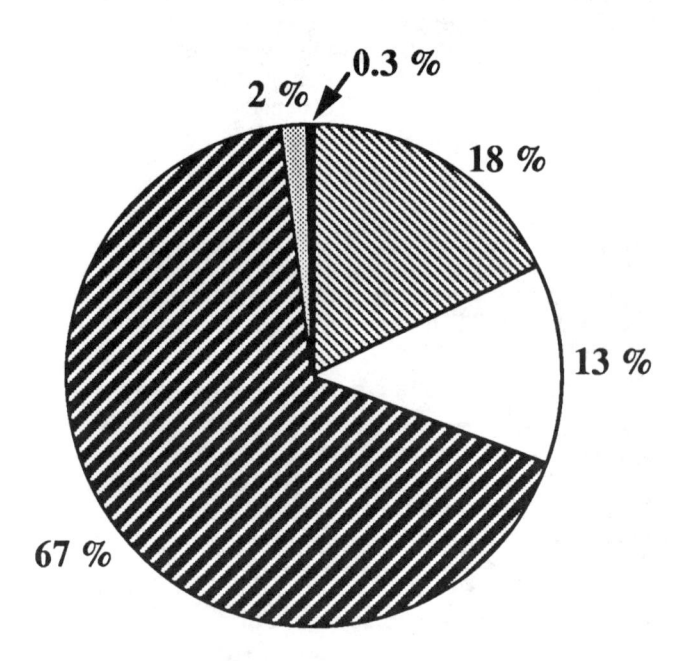

Figure 5: The typical funding sources are indicated for the responding universities in The 1992 NGV Challenge. (Shading as in figure 6.)

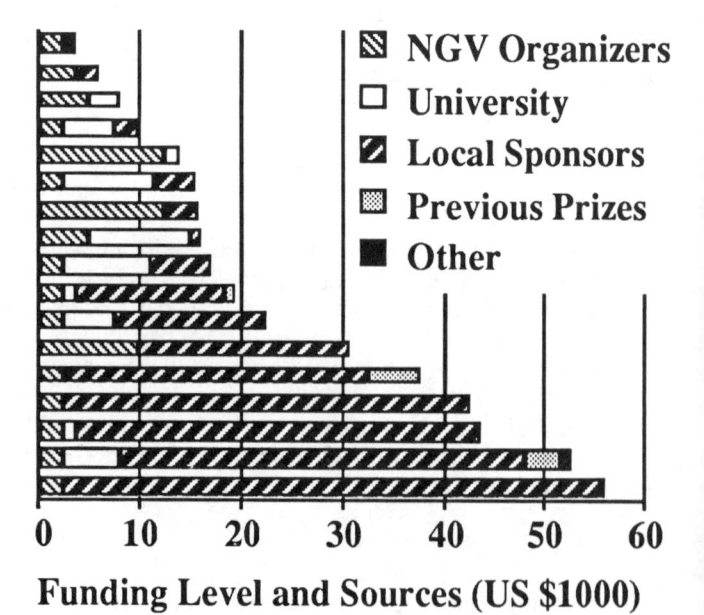

Figure 6: The total funding for responding universities is shown as the total length of the bar. Various sources are identified by the shading of the bar. Bars are ordered from smallest funding level to greatest funding level.

In-kind contributions by equipment suppliers were substantial for many teams, averaging $8,100. Some teams received in-kind contributions amounting to as much as $20,000. Several respondents noted that since some parts donated by industry were prototypes, attaching a dollar value to them was very difficult.

In spite of the wide range of funding levels at the different universities, the faculty advisors were generally satisfied with their funding level. Forty-one percent believed that the funding level was adequate and did not need more funds. Twenty-nine percent could have used more funds but felt that the level was adequate. Twenty-four percent had inadequate funding. Only six percent (one university) had more than enough funding.

FACULTY ADVISOR COMMENTS

Perhaps the most interesting questions in the survey were those seeking open-ended responses from faculty advisors. From the standpoint of using the project as a learning experience for the students, the faculty advisors found the project challenging. One difficulty was integrating a project that would be in existence for only a few years into the standard curriculum. Another problem was how to effectively channel all of the practical knowledge necessary for a design project as large as a vehicle design to the students without getting into the mode of "auto mechanics" instead of engineering. On the other hand, several faculty advisors found it very satisfying to be able to work one-on-one with students and help them reduce what they have learned into practice. Faculty advisors were also pleased to help students learn how to work together and get a taste of how engineering works in industry.

In spite of the rewards of working closely with talented students, several faculty advisors commented on the practical problems of working with undergraduates on this project. Often students were overconfident of what they could realistically accomplish. They underestimated the time required for a particular aspect of the work. Keeping the students focused and moving forward on the project was difficult. Faculty advisors also commented on the challenges of keeping students motivated, encouraging proper student follow-up with sponsors, and ensuring accurate results of testing done by the students. In addition, faculty advisors found that they had to help students cope with the emotional highs and lows that go along with working on this sort of project. Finally, advisors had trouble finding enough time to put in on the project themselves, given their other teaching and research responsibilities.

One of the challenges that several faculty advisors mentioned was that of organizing the students' work on the project. Developing a team approach with the different personalities, motivations, and time commitments of the students involved was not easy. Coordinating and communicating among students was also a problem because undergraduates have different schedules and do not have a work phone as they would in industry. All of these factors made keeping the students on schedule difficult. Finally, controlling the limited financial resources with numerous students moving in many directions was challenging.

Several faculty advisors were disappointed by the lack of support for the project from their universities. Faculty advisors concluded that automotive design projects such as The Natural Gas Vehicle Challenge provide students with an unparalleled, first-hand experience in design. In addition such projects stimulate publicity for the university, enhance relations with industry, and may lead to further research opportunities. Unfortunately, few universities recognized the value of the projects and the commitment that faculty advisors make when they advise such projects. Specifically, faculty advisors suggested that universities recognize the educational and research value of the advisor's work on design projects in promotions and tenure. They also suggested that universities provide release-time from classroom teaching for advisors to work on student design projects. Further, faculty advisors suggested that universities provide better automotive engineering facilities. Very few universities had the chassis dynamometers and sophisticated emissions testing facilities necessary for this type of design competition. Finally, faculty advisors urged that their universities contribute more in financial support of the project to encourage such projects as a format for design education.

CONCLUSIONS

The Natural Gas Vehicle Challenge and similar large-scale automotive design projects offer engineering students an outstanding opportunity to learn about the realities of engineering design. The projects provide students with practical experience that can be incorporated into the engineering curriculum via the capstone design course or an automotive engineering course. Including the project in a course offers the advantages of structuring the students approach to the vehicle design and providing the faculty advisor with a format in which to guide to vehicle development so that students obtain the maximum learning from the experience. Using the project as part of a design course also offers the faculty the opportunity to include advising the project as part of their regular teaching load.

Advising a large-scale automotive design project can lead to research opportunities. For faculty advisors already doing research in automotive engineering the project can provide the means to expand their research. For faculty advisors who are interested in pursuing the wide range of opportunities in automotive engineering and alternate fuels, the project can provide a vehicle to learn about the field and make contacts in the industry. These industrial contacts can also be used to provide real engineering problems that can be used as projects in capstone design classes.

In advising a project like The Natural Gas Vehicle Challenge the faculty advisor takes on a challenging responsibility requiring a substantial commitment of time. But the reward of advising an automotive design project is that a professor can work shoulder to shoulder with students helping them learn about engineering design from first-hand experience.

References

1. "Developing Dedicated Natural Gas Vehicle Technology: 1991 Natural Gas Vehicle Challenge," SAE Publication SP-894, 1992.
2. "Enhancing Natural Gas Vehicle Technology: 1992 Natural Gas Vehicle Challenge," SAE Publication SP-929, 1992.

940557

Analysis of Data from Electric and Hybrid Electric Vehicle Student Competitions

Keith B. Wipke
National Renewable Energy Lab.

Nicole Hill and Robert P. Larsen
Argonne National Lab.

ABSTRACT

The U.S. Department of Energy sponsored several student engineering competitions in 1993 that provided useful information on electric and hybrid electric vehicles. The electrical energy usage from these competitions has been recorded with a custom-built digital meter installed in every vehicle and used under controlled conditions. When combined with other factors, such as vehicle mass, speed, distance traveled, battery type, and type of components, this information provides useful insight into the performance characteristics of electrics and hybrids. All the vehicles tested were either electric vehicles or hybrid vehicles in electric-only mode, and had an average energy economy of 7.0 km/kWh. Based on the performance of the "ground-up" hybrid electric vehicles in the 1993 Hybrid Electric Vehicle Challenge, data revealed a 1 km/kWh energy economy benefit for every 133 kg decrease in vehicle mass. By running all the electric vehicles at a competition in Atlanta at several different constant speeds, the effects of rolling resistance and aerodynamic drag were evaluated. On average, these vehicles were 32% more energy efficient at 40 km/h than at 72 km/h. The results of the competition data analysis confirm that these engineering competitions not only provide an educational experience for the students, but also show technology performance and improvements in electric and hybrid vehicles by setting benchmarks and revealing trends.

INTRODUCTION

The U.S. Department of Energy (DOE), through the Center for Transportation Research at Argonne National Laboratory (ANL), has sponsored Engineering Research Competitions (ERCs) across the nation since 1987. These competitions have involved high schools, vocational schools, community colleges, and universities. Each year the level of student involvement increases along with the technical objectives of the competitions. In 1993, DOE sponsored one hybrid electric vehicle (HEV) and three electric vehicle (EV) competitions, the subjects of this paper, a dedicated natural gas vehicle competition, and the alcohol-fueled classes in Formula SAE and SAE Supermileage. Over 100 different schools participated in these competitions. This year the EV competitions covered the Southwest, Southeast, and Northeast regions of the United States. Teams from all over North America, including two teams from Canada, participated in the HEV competition in Dearborn, Michigan.

The competitions have been a cooperative effort between government, industry, and academia. These competitions have increased awareness of alternative transportation technologies and laid the foundation for collecting data on these technologies. This year, DOE implemented the use of kilowatt-hour meters for data collection during the competitions. These data acquisition meters were donated by DOE to each of the student teams participating in the competitions as part of an ongoing effort to encourage the schools to continue research in the areas of alternative transportation technologies. The student teams were supplied the meters, shunts, and batteries for the meters. Most teams installed the simple, cost-effective meters before the competition and were ready to collect data before it started. This enabled some teams to learn how to drive their vehicles efficiently prior to the start of the events, allowing them the opportunity to perform better in the overall competition.

State-of-the-art technology is demonstrated in the competitions with events highlighting range, acceleration, and efficiency. The basics of the practicality, design, manufacturability, associated costs, and ergonomics of the vehicles were also covered in the competitions, which included oral design presentations. Though the majority of the vehicles used lead-acid batteries and DC motors and controllers, this commercially available technology is showing improvements in performance and reliability. Some of the emerging technologies were revealed at the American Tour de Sol and the 1993 HEV Challenge, displaying promise for the future. The results from the American Tour de Sol, along with detailed event descriptions, have already been presented at a recent Northeast Sustainable Energy Association conference [1], so only the performance of a few representative vehicles will be discussed here.

While the vehicles themselves are normally the focus of these competitions, the charging facilities for electric and hybrid vehicles are also being developed at a fast pace.

Although still in the developmental stages, the charging facilities used in these competitions included such new devices as the 208 V, 30 A, single-phase individual charging meter, developed by Detroit Edison.

BRIEF DESCRIPTION OF THE ENERGY METER

Electrical efficiency data collected in earlier DOE-sponsored EV competitions used an approach developed for the American Tour de Sol in 1991 employing multimeters and a clamp-on current probe to measure average battery pack voltage and average current [1]. A more accurate and permanent probe to record energy efficiency was desired, leading to the development of the energy usage meter used in the 1993 competitions. A kilowatt-hour meter was custom built for ERCs by Cruising Equipment Company according to ANL specifications [2].

The data acquisition unit measured 11.3 cm x 10.1 cm x 4.0 cm, making it suitable for installation in dashboards or elsewhere in the passenger compartment (see Figure 1). The unit recorded the elapsed time and measured two input signals: battery pack voltage and electric current to or from the battery pack. It used this information to calculate the amount of energy running into the batteries (charging and regenerative braking) or out of the batteries (power used to propel vehicle) through a numeric integration. The voltage and current were sampled at 1.2 kHz with averages computed every 128 samples for the energy calculation. The voltage, electrical current, and net energy used were also sent every second to the RS-232 connector through the external port. The current was measured through a shunt with the ratio of 500 A to 50 mV. The voltage was measured by reducing it with a voltage divider on the bus connector [3].

Figure 1. Photograph of the energy meter, Phoenix.

The unit had a numeric LCD to display the voltage, current, or net energy used by means of a three-way switch. Capturing the temporal data required a laptop computer with appropriate software that stored all the information from the serial port. Because of the expense and weight of portable computers, only a few schools used an on-board computer to capture this information. Figure 2 shows temporal data from a 25-lap run by Cortez High School at the Phoenix competition as an example. Notice how the current was used in short bursts, except for the end of the run when the driver had the vehicle under full power for the final five laps. Although the instantaneous power varied continuously, the net energy used (with the negative indicating energy extracted from the batteries) appeared to increase linearly because of the small scale on the graph. This means that the time-averaged power was almost constant at 22 kW. Another interesting feature of this graph is that it shows the effect of the electrical load on the battery pack voltage. This particular vehicle started out with an open-circuit voltage of 102 V, and the battery pack voltage was reduced to 82 V when the electrical load was applied. The graph also clearly shows the battery pack voltage increasing with decreased electrical load, as the vehicle exited the race track and went from a low of 70 V back up to 90 V.

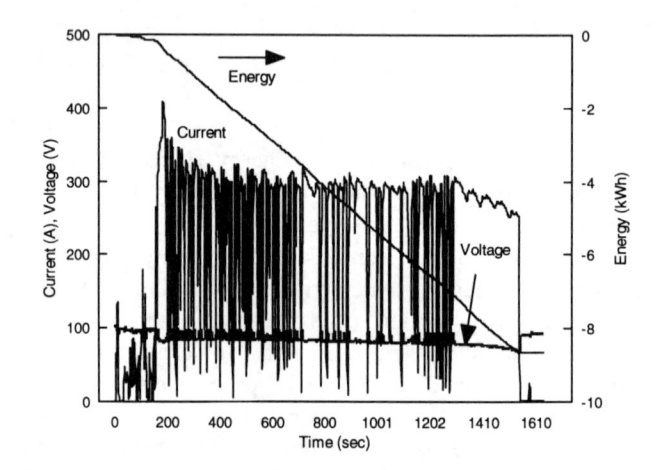

Figure 2. Sample temporal data taken from Cortez High School, Phoenix.

DESCRIPTION OF THE COMPETITIONS: PHOENIX, ATLANTA, DEARBORN, AND AMERICAN TOUR DE SOL

1993 APS Solar & Electric 500, Phoenix

Phoenix was the setting for the 1993 APS Solar & Electric 500, the first of the year's ERCs sponsored by DOE, held March 5 and 6. One New Mexico and 25 Arizona high school teams converted gasoline cars and trucks into EVs. The DOE student engineering competitions use static and dynamic events to educate students in currently available electric vehicle technology.

The two-day competition included static events, such as the Oral Presentations and Design Event, and dynamic events including the Energy Efficiency Event, Range Event, and Acceleration/Braking Event. Both types of events were incorporated to display the full range of students' involvement with their vehicle projects. The static events allowed the students the opportunity to work on their communication and analytical skills, while the dynamic events allowed them to test their vehicle designs and discover the satisfaction of seeing a project to its completion.

Vehicles were built to meet the technical specifications of the 1993 Phoenix Solar & Electric 500 Rules, which focused on the basics of electric vehicle technology for the high school competition class. The vehicles were limited to 96 V lead-acid battery packs, and in most cases used DC motors and controllers donated by General Electric.

The high school vehicles were not separated by vehicle type or weight. For example, half-ton pickup trucks competed against VW Rabbits. To help compensate for the weight difference of the vehicles, a rule of thumb based on experimental results was applied to the points available in the Efficiency Event. For every 10% decrease in vehicle weight, a 7% increase in efficiency was expected. A recent article on weight-cutting efforts claims that, in practice, a 10% weight reduction results in a 3% to 6% fuel economy benefit for conventional vehicles [4]. The weight factor for each vehicle was calculated as

$$\text{Weight Factor} = \left[\frac{\text{Vehicle Weight}}{\text{Average Vehicle Weight}}\right] \times 0.7 + 0.3 \quad (1)$$

Each vehicle's energy economy was divided by this factor, which could be greater or less than one. The average competition vehicle weight was used as the baseline for this equation in the Efficiency Event. The vehicle data used for this weight factor is listed in Table A1 of the Appendix.

After each vehicle successfully completed the safety and technical inspections, the next step was the Acceleration/Braking Event. The acceleration times indicated performance characteristics of the vehicles. The event itself covered 0.40 km, subdivided into a 0.20 km acceleration run and a 0.20 km braking distance. The time required to complete the total distance was used to score this event.

The Range Event measured the farthest distance traveled by a vehicle at a constant speed in 1 hour. A pace car led the single-file line of vehicles around the track at a speed of 88.5 km/h for the entire event, with no passing allowed. The actual average speed over the hour-long event was 81 km/hr because the first few laps were used to get the vehicles up to speed. As vehicles lost energy and slowed below the required speed, they had to leave the track. The number of laps, elapsed time, and energy used was recorded for the vehicles as they left the track. Unfortunately, some confusion with the flagmen occurred when gaps between the vehicles were formed. As a result, two vehicles were flagged off prematurely and were not allowed to reenter the track. The range of these two vehicles was adjusted to compensate for their premature exit on the basis of their energy readings and those of the other vehicles which ran until their energy was depleted. For future competitions, this situation could be alleviated by using a radar gun, requiring accurate speedometers, and allowing passing under controlled conditions.

The Efficiency Event was scored by using the raw data from the meters and the distance traveled during the Range Event to calculate the distance per amount of energy used. The vehicle that demonstrated the most efficient use of energy, after applying the weight factor discussed earlier, won the Efficiency Event.

The Phoenix competition provided a testing ground for the meters and determined the best overall performing vehicle. Despite a short delivery schedule and installation period for the meters, the high schools did an outstanding job incorporating them into the vehicles. The information from the meters proved invaluable for the other electric vehicle competitions that followed.

1993 Clean Air Grand Prix, Atlanta

The Atlanta Clean Air Grand Prix, held in conjunction with the Clean Air Exposition on May 13, 14, and 15 was a collegiate EV competition. Fourteen teams of universities, community colleges, and technical schools from the Southeast were invited to participate, making it the first electric vehicle competition held in that region of the United States. Of these fourteen schools, three were historically black colleges. Most of the teams had from January until May to design and convert their vehicles to run on electric power. Because of the short amount of preparation time, only ten of the teams competed in the Grand Prix. The Clean Air Vehicle Association (CAVA) put together the Clean Air Grand Prix to educate the public and the students on electric vehicles, as well as to add flair to the exposition.

CAVA worked with a number of sponsors to have most of the items donated to all the teams, including the vehicles, batteries, tires, motors, controllers, and the energy meters. Trojan Battery Company provided the batteries, General Electric provided the motors, controllers, and technical support for the teams, and Arena Auto Auctions donated suitable cars to the teams. CAVA developed a set of rules and technical specifications based on the Sports Car Club of America's touring rules. All teams were limited to the 120 V battery packs and DC motors and controllers. While there were no pickup trucks in this competition, the donated vehicles varied in model, style, and weight (see Table A2 for detailed Atlanta competition vehicle data). Because the teams had no say in the vehicle they received, normalizing the vehicle weights was important in scoring the events affected by this factor. The variation in the weight of the vehicles was compensated for by using Equation 1.

The structure of the Atlanta competition was similar to the one in Phoenix with a layout that included acceleration, range, electrical efficiency, design review, and oral presentations. However, ANL personnel worked closely with the event organizers and had input into the type of data collected and the manner in which it was collected. For example, the Acceleration Event was modified to include a five-lap solo event, and the Efficiency Event was separated from the Range Event. This allowed for data collection at three separate constant speeds, with additional control over collection methods.

The Acceleration/Solo Event added a five-lap solo run to the 0.20 km acceleration run. The solo portion tested the performance of each vehicle and allowed drivers time alone on the track to test their vehicles without the complication of passing. The 0.20 km acceleration run was held on a 1.42 km road track at the Atlanta Motor Speedway. In addition to the track times, kilowatt-hour readings were recorded for each vehicle. This provided a history of energy usage for each vehicle, to be used as a backup in case any problems with the meters occurred during other events.

The Range Event used the same road course as the Acceleration/Solo Event. The layout of the course did not lend itself to the type of pace car setup used in Phoenix. Instead, the vehicles had a time limit of 2 hours and a minimum lap speed of 40 km/h. This gave the teams greater flexibility with their individual driving strategies. Passing was allowed only on the straightaways, and vehicles were allowed to pull off the track and reenter at any time during the 2 hour limit.

The Efficiency Event measured the energy consumption of the vehicles at three distinct, constant speeds. A pace car led the vehicles around the 2.4 km oval for a total of five laps at 72 km/h and 56 km/h and three laps at 40 km/h. The meters were read just before and just after each group of laps was driven.

1993 Hybrid Electric Vehicle Challenge, Dearborn

The 1993 HEV Challenge, held June 1-5, was the first in a series of competitions focusing on the emerging technologies associated with hybrid electric vehicles. These vehicles combine the best features of the liquid-fuel-powered vehicle and the electric vehicle while offering the range and performance provided by conventional vehicles. HEVs also provide an emissions-free option for zero emissions zones. Thirty colleges and universities designed and built parallel or series hybrid vehicles for the competition. The teams had the choice of building a vehicle from the ground up (Ground-Up Class) and receiving $10,000 in seed money, or converting a donated 1992 Ford Escort Wagon to a hybrid vehicle (Escort Conversion Class). Eighteen of the teams selected chose to convert Escorts while the remaining 12 teams built vehicles from the ground up (see Table A3 and A4 for details on the hybrid strategy and components used for the Ground-Up and Escort Conversion classes, respectively). The teams participated in a five-day competition that covered qualifying (See Figure 3), emissions testing, acceleration, range, vehicle efficiency, a simulated commuter event, oral presentations, design reviews, and cost assessments. All aspects of vehicle design, construction, and performance were judged and scored. Twenty-six out of the 30 HEVs completed that portion of the Challenge.

One major goal of the HEV Challenge was to explore efficient vehicle propulsion systems. Events were specifically designed to determine the overall efficiency, electrical efficiency, and alternative power unit (APU) efficiency of the vehicles by measuring the fuel consumption and energy usage throughout the competition. The Electrical Efficiency Event, for example, required the teams to meet a minimum electric-only or zero emission vehicle (ZEV) range requirement of 32.2 km while determining which electrical system functioned most efficiently without using the APU. The kilowatt-hour meters, charging stations, and ZEV events were used for this purpose [5]. The Range Event was divided into four segments: a 32.2 km portion around a track at the Dearborn Proving Grounds (DPG), followed by a 129.4 km HEV portion on public streets where the APU could be on, followed by ZEV and/or HEV driving modes on the track at Michigan International Speedway (MIS). Efficient charging schemes and skillful driving were encouraged by measuring the kilometers driven per kilowatt-hour of energy consumed. A loss factor representing energy losses in the vehicle charging systems was applied in computing the electrical efficiency results.

Figure 3. Photograph of the University of Tennessee vehicle at the acceleration qualifying event, Dearborn.

1993 American Tour de Sol

The focus of the American Tour de Sol is educating the general public on alternative renewable energies available for transportation. The first Tour de Sol included solar cars and electric vehicles that were very different in construction and appearance. This year the competition involved over thirty vehicles ranging from solar cars to prototype EVs and included the first HEV to participate in the six day rally (see Table A5 for vehicle data from the Tour de Sol). Tour de Sol started in Boston, MA and ended up in Burlington, VA. The rally included check points and time limitations for each portion of the route, with extra points given for additional miles completed at the end of each leg of the rally. The kilowatt-hours were recorded at the beginning and end of each day, along with the total distance traveled by each vehicle. By the end of the week, all but two teams had their meters working. The teams with AC systems had difficulty with the meter operating properly. The sensitivity of the meter to voltage and current inputs and the electrical noise of the AC motor and controller both contributed to the problems encountered with the data collection.

The American Tour de Sol, being the longest running competition of its kind, has many repeat competitors. While the improvements in EV technology are recorded each year, the difficult terrain presents the opportunity to observe the reliability of these electric vehicles. The majority of the vehicles used lead-acid battery technology and demonstrated a repeatable range of 129 km each day. Many of the vehicles with advanced battery technology (Zinc Bromine and NiCad) displayed ranges of 241 km or more. The American Tour de Sol is a proving ground for many student-based electric vehicles.

RESULTS OF ANALYSIS

Open-Circuit Voltage Drop vs. Distance in Lead-Acid Battery Packs

The open-circuit voltage drop in lead-acid battery packs was measured as a function of the number of kilometers driven by the EVs during the Phoenix competition. In lead-acid batteries, the open-circuit voltage decreases as a function of state-of-charge (SOC). The open-circuit voltage of each vehicle was measured prior to its starting an event and again just after the vehicle had pulled off of the track. As expected, the open-circuit voltage was still changing with time when the vehicles were checked after pulling off the track. This is primarily caused by the batteries recovering from such a rapid discharge, and involves a combination of the batteries cooling down and the chemistry in the batteries returning to equilibrium. For consistency, the voltage was measured immediately after the vehicle stopped moving; however, this was not possible when multiple vehicles exited the track simultaneously.

As shown in Figure 4, the voltage drop was generally larger the farther the vehicle traveled, but there is scatter caused by differences in vehicle mass, vehicle efficiency, initial SOC, rate of discharge, and battery manufacturer. However, the data does indicate the expected trend of increasing battery pack voltage drop with increasing distance traveled. All of the EVs at Phoenix had lead-acid battery packs rated at 96 V, so the data in Figure 4 show a battery pack open-circuit voltage drop of between 3% and 18% (see Table A6 for event data for Phoenix, Atlanta, and Dearborn).

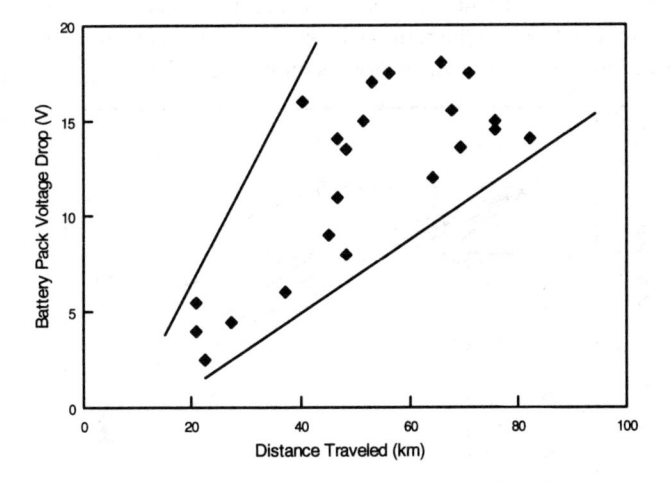

Figure 4. Open-circuit voltage drop in lead-acid battery packs vs. distance, Phoenix.

Effect of Vehicle Mass

At all of the competitions, the vehicles were weighed with four electronic scales, one placed under each wheel, to measure the weight and mass and to determine the mass distribution. Coupling the vehicle mass with its energy economy, defined as distance traveled divided by energy used, provided insight into the significant effect of vehicle mass on electric and hybrid vehicles. At the HEV Challenge in Dearborn, the vehicles which most clearly showed this effect were the Ground-Up hybrids, as these vehicles displayed considerable spread (588 kg) in vehicle mass. Because the Escort Conversion vehicles had a maximum vehicle mass allowed by the rules and regulations (gross vehicle weight rating + 10%), their masses were concentrated around this specification with only a 234 kg spread (see Table 1, noting that only vehicles from the three competitions which competed and provided accurate meter readings are included in this analysis). The difference in the components and designs of the Escort Conversion vehicles proved more significant than the small difference in the vehicles' mass.

Table 1. Vehicle mass

Competition	Min. Mass (kg)	Max. Mass (kg)	Spread (kg)	Avg. (kg)
Dearborn (Ground-Up)	1,062	1,650	588	1,306
Dearborn (Escort Conversion)	1,614	1,848	234	1,695
Atlanta	1,023	1,703	680	1,358
Phoenix	1,307	1,645	338	1,446

Figure 5 shows energy economy plotted against mass for the Ground-Up vehicles at Dearborn during the first ZEV portion of the Range Event. There are also many differences between these vehicles in addition to the mass, such as differing components, body aerodynamics, and tires. However, because the heaviest vehicle had a mass 55% higher than the lightest vehicle, and the vehicles were not traveling at high speeds, the large difference in mass dominates the differences in component efficiencies. A linear interpolation showed a 1 km/kWh increase in energy economy for every 133 kg decrease in vehicle mass. A detailed discussion of the benefits of lightweight hybrid vehicles is given by Lovins, Barnett, and Lovins [6].

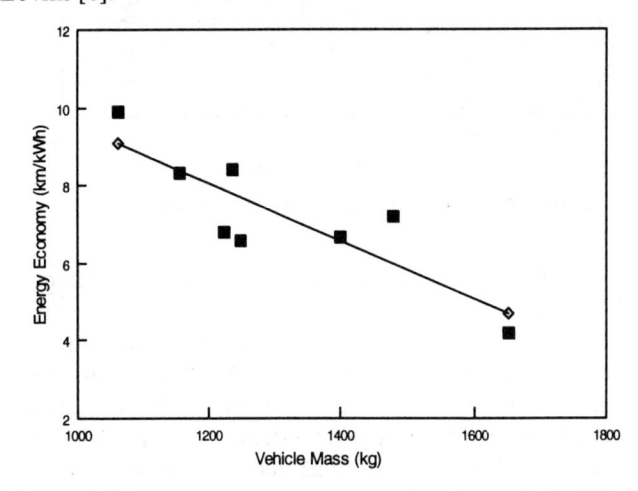

Figure 5. Energy economy vs. mass for Ground-Up HEVs, Dearborn.

As previously discussed in the description of the competitions, the Atlanta competition provided an ideal setting to examine the effect of mass and speed. The three runs at 72, 56, and 40 km/h provided energy usage data for a fixed distance around a 2.4 km track. The difference between the most and least massive electric vehicles at Atlanta was 680 kg. With the exception of the data from one vehicle at 1250 kg, Figure 6 shows a relatively smooth curve, with the lightest vehicles having the highest energy economy and the heaviest vehicles having the lowest energy economy. The three symbols arranged in a vertical line are from the same vehicle at the three different speeds.

All of the Atlanta vehicles were conversions, but the major differences in mass were of similar magnitude to the Ground-Up HEVs and had more effect than other differences in the vehicles. For current conventional U.S. cars, fuel economy is about equally sensitive to reductions in drag and rolling resistance, but is nearly three times as sensitive to reductions in mass [7,6]. Because the Atlanta vehicles were all conversions from conventional vehicles, it is not surprising to see a similar sensitivity to mass displayed in Figure 6.

Another notable feature of the Atlanta competition was that sponsors provided the same components to all vehicles, with the exception of the battery manufacturer sponsor, which allowed the schools to select the battery type most suitable for meeting their goals. Some teams selected batteries which provided optimal power and acceleration, whereas other schools favored increased energy capacity and driving range.

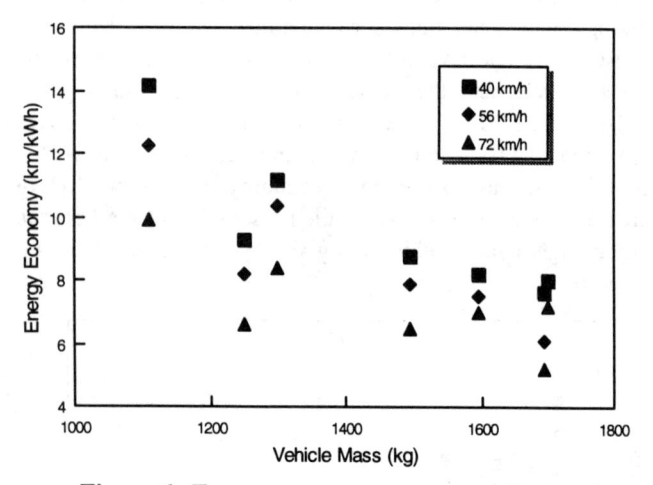

Figure 6. Energy economy vs. mass, Atlanta.

Although the effect of vehicle mass was found to be significant in all three competitions, the data from Phoenix does not indicate this as clearly as the other competitions because of the nature of that event. At Phoenix, all the vehicles had to maintain a constant speed of 88.5 km/h behind a pace vehicle. If a vehicle started lagging behind, it was flagged off of the track. Therefore, the Phoenix data is not as representative of how much energy the vehicles were carrying as it would have been if they had been allowed to drive until their batteries were depleted. Aerodynamic drag also played a larger part in Phoenix because of the relatively high speed (81 km/h) at which the vehicles were driven. Additionally, the difference between the most and least massive vehicles was not as large as in the other two competitions. At Atlanta and Dearborn, however, there was a good correlation between vehicle mass and energy economy, as has already been discussed.

Effect of Vehicle Speed

In addition to showing the effect of vehicle mass, Figure 6 also reveals the effect of different vehicle speeds with the three vertically spaced symbols. It is interesting to note how much more the lightest vehicle was affected by aerodynamic drag than the heaviest vehicle, as judged by the vertical distance between symbols. This is because the lighter vehicle is spending a larger percentage of its power on overcoming aerodynamic losses than it is in overcoming rolling resistance. Therefore, the effect of vehicle speed on energy economy is more pronounced for lighter vehicles than for heavier vehicles for a constant speed. The primary forces involved are

$$F_{total} = F_{roll} + F_{aero} = fmg + \frac{1}{2}\rho C A v^2 \quad (2)$$

where f is the coefficient of rolling resistance, C is the coefficient of aerodynamic drag, A is the frontal area, and ρ is the air density [8]. This equation shows that the rolling resistance is proportional to the vehicle mass (m) while the aerodynamic drag increases by the square of the velocity. When the energy economy is plotted against vehicle speed rather than mass, the effect of the speed on energy economy becomes easier to see, as shown in Figure 7. The energy economy has been normalized at 40 km/h, so that the ratio at 40 km/h is 1, while the ratio at 72 km/h is the energy economy at 72 km/h divided by the energy economy at 40 km/h.

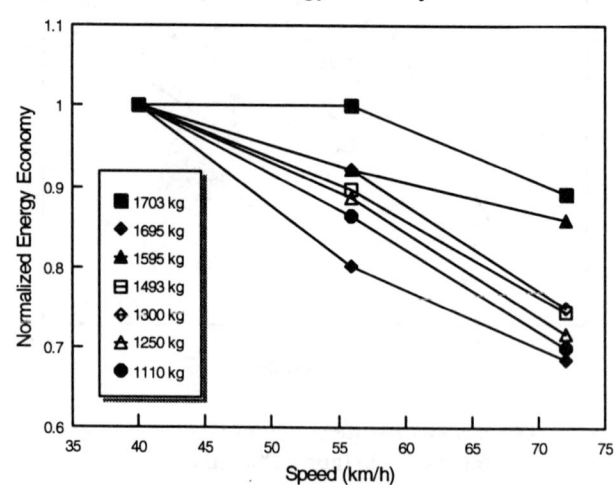

Figure 7. Normalized energy economy vs. speed, Atlanta.

All of the vehicles showed a decrease in energy economy as the vehicle speed was increased, except for the 1703 kg vehicle which stayed the same between 40 and 56 km/h. More importantly, Figure 7 clearly shows how the lighter vehicles experienced a larger percentage decrease in energy economy compared with the heavier vehicles

(30% decrease vs. 11% decrease). The exception was the one outlying vehicle that had a mass of 1695 kg. Therefore, when it comes to highway driving, vehicle designers should consider the aerodynamic factor to be much more important in minimizing energy usage in the new lightweight vehicles than in conventional vehicles. Lovins supports this conclusion, claiming that in higher fuel economy vehicles, aerodynamic drag is more important than mass [6].

Vehicle Driving Range

The driving range of the electric and hybrid vehicles was tested in all three competitions. In Atlanta, the vehicles had 2 hours, could drive at any speed above 40 km/h, and could even stop for a period of time to let their batteries recover. In Phoenix the range was tested under the rigorous requirements of maintaining 88.5 km/h behind a pace vehicle. Most of the teams at Phoenix could have driven much further if they had been able to slow down as their batteries became depleted, or if they had been able to drive at a slower speed over the whole time period. The time allowed at Phoenix was limited to 1 hour, and only one vehicle was able to maintain the required speed at the end of that time. In Dearborn, the vehicles had ZEV (electric only) and HEV (where the APU was switched on) portions of the Range Event, with an overall time period of 5 hours. The data summarized in the Appendix and Figure 8 represents the data from the two ZEV portions of that event only.

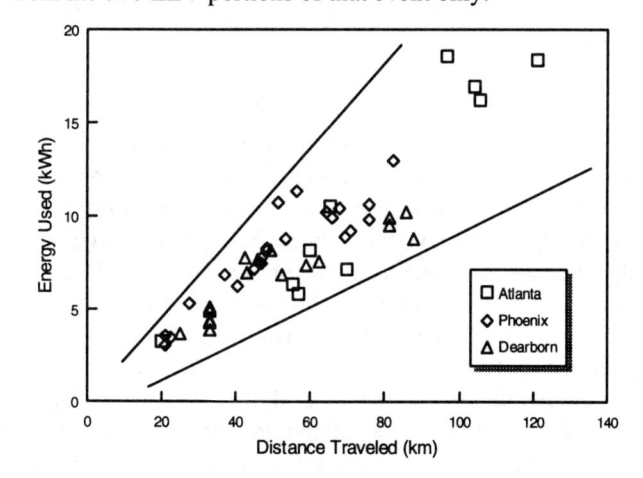

Figure 8. Electric-only range for all 3 competitions.

Figure 8 shows the electrical energy used plotted against distance traveled during the Range Event for each vehicle in the three competitions. The maximum ranges under these conditions for the best vehicles were 121.1 km at Atlanta, 82.0 km at Phoenix, and 87.9 km at Dearborn during the two ZEV portions. Since extended range is one of the advantages of HEVs, it should be noted that the best range achieved at Dearborn, including the APU operational mode, was 303 km. HEVs use two sources of energy, thus complicating energy economy calculations. We will focus on the electric-only energy usage. As will be discussed in the next section, the ratio of distance traveled to energy used provides a measure of a vehicle's energy economy, which is related to how efficiently the vehicle was able to convert its stored energy into moving from one point to another.

Energy Economy and Efficiency

Energy economy, as defined in this paper, is determined by dividing the distance a vehicle travels by the amount of energy it took to travel that distance, expressed in km/kWh. While the term "economy" may seem awkward in the context of energy efficiency, it comes from the term "fuel economy," measured in km/L of fuel, a common measure of how efficient a conventional vehicle is at converting its fuel energy into movement. Some people prefer to use kWh/km so that the numerical values are not mistakenly compared to km/L of fuel by the general public [9]. In any case, the ratio of distance traveled to energy used is the common way of looking at the efficiencies of electric and hybrid vehicles. For data on the energy economy of 30 commercially available electric vehicles, refer to the *Electric Vehicle Directory* [10].

The best energy economy demonstrated during the Phoenix Range Event was 7.75 km/kWh. During the Range Event at Atlanta, the most efficient vehicle had an energy economy of 9.72 km/kWh in 2 hours with an average speed of 35 km/h. Because of the short length of the constant-speed efficiency runs at the three different speeds in Atlanta, their Efficiency Event should not be compared to the other competitions' Efficiency Events which involved much greater distances. However, it should be noted that the same school with the highest energy economy in the Range Event at Atlanta also showed a record high of 14.2 km/kWh at the constant speed of 40 km/h. During the first EV portion of the Range Event at the HEV Challenge, the most efficient Ground-Up hybrid had an energy economy of 9.89 km/kWh, with the most efficient Escort Conversion demonstrating a 9.70 km/kWh energy economy.

Most of the energy economy values obtained from the competitions are on closed-course tracks, bringing into question the correlation of these values with what would occur on public roads under normal operation. One of the strengths of the American Tour de Sol competition is that the entire rally part of the competition takes place on public roads in more than one state. Table A7 compares the University of Central Florida (UCF) vehicle during the Range Event in Atlanta with three American Tour de Sol vehicles on day 4, showing comparable performance from the Kineticar (7.83 km/kWh) with the UCF vehicle (6.56 km/kWh). Two other American Tour de Sol conversion vehicles are also in the table to show that many vehicles performed much better than this at similar speeds.

Although the efficiency numbers generated at the different events are not directly comparable, they establish benchmarks against which existing and future EVs will be judged. Besides being tools for evaluating the development of EVs over the road and on closed courses, they set a standard for which vehicle designers and student engineers can aim. However, besides the energy economy of EV operation, the overall energy economy of EVs will be affected by the efficiency of their charging systems, as will be discussed in the next section.

Energy Charging Loss Factors and Their Effect on Overall Electric Vehicle Efficiency

As part of the charging facilities constructed for the HEV Challenge provided by Detroit Edison, individual charging stations were equipped with meters to measure the amount of AC line energy used by the HEVs to recharge their batteries after a full day of testing. The vehicles had six hours of recharging time, between 11 PM and 5 AM. This energy was recorded and then compared to the reading on the meters in the vehicles which recorded the amount of energy that reached the vehicles' traction batteries. The difference between the line energy and the energy that reached the battery pack determined the charging efficiency of the HEVs, and established a charging loss factor as the ratio of line energy to the energy that entered the battery pack [5]. This loss factor, which was always greater than 1, was used to calculate the total electrical energy used by the HEVs for scoring the Energy Efficiency Event.

Table 2. Loss factors and Charge Efficiencies, Dearborn

Ground-Up Class	Loss Factor	% Effic.
Cal Poly - San Luis Obispo	1.21	82.4
Cornell University	1.67	60.1
Lawrence Technical Univ.	2.93	34.1
Michigan State Univ.	1.75	57.3
Univ. Cal - Davis	2.16	46.3
Univ. of Tulsa	3.23	31
Average	2.61	38.3
Conversion Class		
Cal State - Northridge	2.14	46.8
Colorado School of Mines	2.2	45.4
Colorado State Univ.	3.23	31
Concordia Univ.	3.23	31
Seattle Univ.	1.09	91.5
Stanford Univ.	1.48	67.5
U.S. Naval Academy	3.23	31
Univ. of Alberta	2.3	43.5
Univ. Illinois	2.53	39.6
Weber State Univ.	1.68	59.6
Average	2.07	48.3

The loss factors illustrated in Table 2 show that the efficiencies of the electric vehicle charging systems in use at the HEV Challenge cover a wide range from a modest 31% to a respectable 91%. Vehicles for which a loss factor was not available used the average loss factor from all vehicles, and are not shown in Table 2. With a median value of 50% and an average value of 38% for the Ground-Up class and a median of 50% and an average of 48% for the Conversion class, it is clear that charging inefficiencies detract from the otherwise impressive on-road electrical energy efficiency of EVs and HEVs. If these results are representative of typical charging technologies employed today, improving charging efficiencies is an area where more attention should be focused. The results show that on-board vehicle recharging systems with over 90% efficiency already are available; the selection of more efficient charging equipment and strategies should be adopted by EV and HEV designers. In addition, these results show that in order to fully understand the energy utilization of EVs and HEVs, their total electricity usage, including charging loss factors, needs to be incorporated into any assessment of their economic and environmental costs.

FUTURE ACTIVITIES

DOE will continue to support ERCs and collect data from them in 1994. Data on more types of EVs, including pre-production prototypes from major manufacturers, will be collected with plans to perform dynamometer efficiency testing as well as over-the-road and closed-course testing of EVs. To better understand HEVs, DOE, through the National Renewable Energy Laboratory, will be developing more sophisticated onboard data acquisition systems for the 1994 HEV Challenge. Additional information on HEV operating characteristics is planned to be collected to better characterize the performance levels and help determine the most promising components and vehicle configurations. The results from next year's competitions will also enable meaningful comparisons to be made with the performance data collected in 1993.

CONCLUSIONS

The electrical energy from the 1993 electric and hybrid vehicle competitions was measured with a custom-built digital meter installed in every vehicle. As expected, the mass of the vehicle has been shown to have a significant effect on vehicle energy efficiency. Data from the Atlanta competition has also demonstrated that mass becomes even more critical in designing new vehicles as the vehicles become more energy efficient. The average energy economy for the three competitions analyzed in this paper was 7.0 km/kWh, with the highest energy economy recorded during any event being 14.2 km/kWh. DOE will continue to sponsor engineering research competitions for high schools, vocational schools, community colleges, and universities for the purposes of educating students in the advanced transportation field and collecting data to demonstrate the current capabilities of electric and hybrid vehicles.

REFERENCES

1. Wills, R., "Technical Report for 1993 American Tour de Sol," Northeast Sustainable Energy Association Conference, Greenfield, MA, 1993.

2. Proctor, R., "Catching the Dream," *Home Power* #37, October/November 1993.

3. Cruising Equipment Co., "Installation Manual: KiloWatt-Hour+ Meter," Seattle, WA, February 1993.

4. Fleming, A., "Putting Fuel on the Line, Weight Reduction is Key to Achieving Tomorrow's Rules for Fuel Economy," *Automotive News*, November 29, 1993.

5. Ford Motor Company, *1993 Hybrid Electric Vehicle Challenge, Rules and Regulations*, Ford Motor Company, Dearborn, MI, June 1993.

6. Lovins, A. B., Barnett, J. W., Lovins, L. H., "Supercars, the Coming Light-Vehicle Revolution," European Council for an Energy-Efficient Economy, Rungstedgard, Denmark, June 1993.

7. Office of Technology Assessment, *Improving Automobile Fuel Economy: New Standards, New Approaches*, OTA-E-504, U.S. Congress, Washington DC, 1991.

8. Adler, U., *Automotive Handbook*, 2nd Edition, Robert Bosch GmbH, Federal Republic of Germany, 1986.

9. Brasch, P., "A Precision DC Energy Monitor for Electric Vehicles," WESCON/93 IEEE Conference Proceedings, 1993.

10. Centre for Analysis and Dissemination of Demonstrated Energy Technologies, *Electric Vehicle Directory*, Electric Vehicle Progress, New York, NY, October 1991.

APPENDIX

Table A1. 1993 APS Solar & Electric 500 Vehicle Data, Phoenix

School	Mass (kg)	Vehicle Make	Model/Year	Motor Manufact.	Controller Manufact.	Battery Manufact.
Agua Fria	1,290	Volkswagen	Rabbit/1982	GE	GE	U.S. Batteries
Cactus	1,487	Datsun	240/1974	GE	GE	Trojan
Camelback	1,645	Chevrolet	S10 Pickup/1985	GE	GE	U.S. Batteries
Carl Hayden #13	1,459	Toyota	Corolla/1978	GE	GE	Exide
Carl Hayden #36	1,432	Volkswagen	Rabbit/1980	GE	Curtis	Exide
Carl Hayden #69	1,349	Volkswagen	Dasher/1979	GE	GE	Exide
Chapparral	1,481	Chevrolet	Corvair/1965	GE	GE	Trojan
Cortez	1,415	Ford	Escort/1982	GE	GE	Trojan
East Valley Inst.	1,481	Chevrolet	LUV Pickup/1979	GE	GE	Douglas
Farmington	1,471	Datsun	280Z/1976	GE	GE	Caterpillar
Holbrook	1,367	Ford	Festiva/1988	GE	GE	Trojan
Marcos de Niza #93	1,509	Ford	Courier Pickup/1974	GE	GE	Trojan
Marcos de Niza #94	1,451	Datsun	Pickup/1978	Curtis	Curtis	Trojan
McClintock	1,323	Ford	Mustang/1980	GE	GE	Trojan
Mountain View	1,345	Chevrolet	Cavalier/1984	Adv. DC	Curtis	Trojan
North	>2,300	Dodge	Pickup(1/2-ton)/1980	GE	GE	Enerdyne
Page	1,462	Ford	Escort/1984	GE	GE	Trojan
Palo Verde	1,517	Ford	Courier Pickup/1981	GE	GE	Trojan
Paradise Valley	1,464	Ford	Tempo/1988	Adv. DC	Curtis	Champion
Shadow Mountain	1,307	Chevrolet	Chevette/1979	GE	GE	Trojan
Snowflake	1,196	Honda	Civic/1982	GE	GE	Power Battery
South Mountain	1,426	Chevrolet	Citation/1980	GE	GE	GNB
St. Johns	1,334	Chevrolet	Chevette/1977	Adv. DC	Curtis	Trojan
Sunnyside	1,533	Chevrolet	LUV Pickup/1978	Adv. DC	Curtis	Trojan
Window Rock	1,409	Ford	Escort/1983	GE	GE	Douglas

Table A2. 1993 Clean Air Grand Prix Vehicle Data, Atlanta

School	Mass (kg)	Make	Model	Battery Model	Bus Voltage	No. of Batt.	Cell Voltage
Alcorn State	1,082	VW	Rabbit	DC-78	120	10	12
Berea College	1,329	Ford	EXP	5SHP	120	10	12
Clemson	1,493	VW	Rabbit	T-145	120	20	6
Daytona Beach C.C.	1,250	Chevy	Chevette	5SHP	126	10	12
Duke University	1,023	VW	Karmann-Ghia	TH19	96	8	12
Fort Valley State College	1,110	Honda	Civic	27TMH	108	9	12
Kentucky Adv. Tech	1,300	Hyundai	Excel GLS	TMH-27	120	10	12
Kentucky Tech	1,595	Chevy	Chevette	T-145	120	20	6
Louisiana Tech	1,695	VW	Rabbit	J305	120	20	6
Univ. of Central Florida	1,703	Ford	Mercury Lynx	T-145	120	20	6

Table A3. Ground-Up 1993 HEV Challenge Vehicle Data, Dearborn

School	Mass (kg)	Fuel Type	HEV Strategy	Battery Type	APU	Electric Motor	Controller	Battery Charger
Cal. Poly., Pomona	1,248	E100	Series	Pb-Acid	Geo Metro	Adv. DC	n/a	student built
Cal. Poly., San Luis Obispo	1,223	Gasoline	Parallel	Pb-Acid	Geo Metro	Solectria	Solectria	student built
Cornell University	1,155	M85	Series	Pb-Acid	Briggs & S.	Solectria	Solectria	Lester Elect.
Lawrence Tech. Univ.	1,650	Gasoline	Parallel	Pb-Acid	Geo Metro	Magnetek	Square	K & W Engr.
Michigan State Univ.	1,478	Gasoline	Series	NiMH	Geo Metro	GE	Magnetek	n/a
New York Inst. of Tech.	n/a	M85	Series	Pb-Acid	Kawasaki	n/a	n/a	Solar Car Co.
UC Davis	1,062	Gasoline	Parallel	NiCad	Briggs & S.	Unique	Unique	Solectria
UC Santa Barbara	1,401	E100	Parallel	Pb-Acid	Suzuki	Solectria	Solectria	n/a
Univ. of Idaho/Washington	1,983	Gasoline	Series	Pb-Acid	Kohler	AC Prop.	AC Prop.	AC Prop.
University of Tennessee	1,233	Gasoline	Series	Pb-Acid	Kohler	Unique	Motorola	Goodall
Univ. of Texas, Arlington	787	M85	Parallel	NiCad	Honda	Solectria	Solectria	Solectria
University of Tulsa	1,741	Gasoline	Series	Pb-Acid	Honda	Baldor	Baldor	student built

Table A4. Escort Conversion 1993 HEV Challenge Vehicle Data, Dearborn

School	Mass (kg)	Fuel Type	HEV Strategy	Battery Type	APU	Electric Motor	Controller	Battery Charger
Cal. State, Northridge	1,489	Gasoline	Series	Pb-Acid	Kawasaki	Unique	Unique	K & W Engr.
Colorado School of Mines	1,614	E100	Series	NiCad	Suzuki	Adv. DC	Curtis	ByCan
Colorado State Univ.	1,722	Gasoline	Parallel	Pb-Acid	Kawasaki	Unique	Unique	Good-All Elec
Concordia University	1,671	Gasoline	Both	Pb-Acid	Briggs & S.	Adv. DC	Curtis	n/a
Jordan College Energy Inst.	1,694	Gasoline	Parallel	Pb-Acid	Kawasaki	Adv. DC	Curtis	student built
Pennsylvania State	1,966	Gasoline	Parallel	Pb-Acid	Geo Metro	Solectria	Solectria	Good-All
Seattle University	1,715	Gasoline	Parallel	Pb-Acid	Geo Metro	Unique	Unique	Good-All
Stanford University	1,660	Gasoline	Series	Ni-Cad	Honda	Stanford	FMC	Solectria
Texas Tech. University	1,824	E100	Parallel	Pb-Acid	Kawasaki	Solectria	custom	Solar Car Co.
US Naval Academy	1,717	Gasoline	Series	Pb-Acid	Briggs & S.	GE	Curtis	New Concepts
University of Alberta	1,633	Gasoline	Parallel	NiCad	Suzuki	Solectria	Solectria	student built
UC Irvine	1,448	M85	Parallel	Pb-Acid	Suzuki	Electra-Gear	Emerson	Lester Elec/
University of Illinois	1,643	E100	Series	Pb-Acid	Kawasaki	Magnetek	Magnetek	n/a
University of Wisconsin	1,719	Gasoline	Series	Pb-Acid	Kohler	Electric App.	Indramat	n/a
Washington Univ., St. Louis	n/a	E100	Parallel	Pb-Acid	Briggs & S.	Adv. DC	custom	n/a
Wayne State University	1,848	Gasoline	Parallel	Pb-Acid	Ford Escort	Garret	GE	n/a
Weber State University	1,725	Gasoline	Parallel	Pb-Acid	Ford Escort	Adv. DC	Curtis	Indust. Batt.
West Virginia University	1,652	M85	Series	Pb-Acid	Kawasaki	Adv. DC	Curtis	Cybertronics

Table A5. 1993 American Tour de Sol Vehicle Data

Vehicle Name	kg	GU/Conv	Battery Type	Battery Capacity (kWh)	Motor Type	Regen. Braking
Aztec	363	GU	Lead Acid	6.8	DC Brushless	Y
C-M Sunpacer	464	GU	Lead Acid	9	DC	N
Chevy S-10 pjt. E	1,858	Conv.	Lead Acid	27.6	DC	-
Delto Fiero SE	1,533	Conv.	Lead Acid	10.6	DC	Y
Electric Jewel	1,272	Conv.	Lead Acid	17.4	DC Series	N
Electric Lizzie	989	GU	Lead Acid	1.6	DC Series Wound	N
Electric Taxi	1,312	Conv.	Lead Acid	22	DC	N
Envirocycle	242	Conv.	Lead Acid	0.94	DC Series Wound	N
Envirocycle II	751	Conv.	Lead Acid	1.09	DC Series	N
Genesis I	1,683	Conv.	Lead Acid	11.9	DC	Y
Kineticar	1,529	Conv.	Lead Acid	22.2	DC Series	Y
Potential Difference	1,186	Conv.	Lead Acid	18	DC Series	Y
Rham Rod	1,457	Conv.	Lead Acid	17.4	DC	Y
Solar Bolt	1,312	Conv.	Lead Acid	22	DC	N
Solar Bullet	557	GU	Lead Acid	6.8	DC Wire Wound	N
Solar Flair	1,478	Conv.	Lead Acid	22	DC	N
SUNGO	595	GU	Lead Acid	7.2	DC Brushless	Y
T-Star	1,125	Conv.	Zinc Bromine	21.6	AC Induction	Y
Viking 21	858	GU	NiCad	5.7	Unique Mobility	Y

Table A6. 1993 Event Data Atlanta, Phoenix, and Dearborn

Atlanta: Clean Air Grand Prix, May 1993

Efficiency Event (const. speed)

School	Mass. (kg)	72 km/h Energy (kWh)	Distance (km)	Effic. km/kWh	56 km/h Energy (kWh)	Distance (km)	Effic. km/kWh	40 km/h Energy (kWh)	Distance (km)	Effic. km/kWh	Range Event (2 hours, any speed) Energy (kWh)	Dist. (km)	Effic. km/kWh	Solo Event (shortest time) Energy (kWh)	Dist. (km)	Effic. km/kWh
Alcorn State	1082	-----	-----	-----	-----	-----	-----	-----	-----	-----	3.28	20.0	6.08	-----	-----	-----
Berea College	1329	-----	-----	-----	-----	-----	-----	-----	-----	-----	10.50	65.6	6.24	2.63	8.6	3.25
Clemson	1493	1.11	7.2	6.53	0.92	7.2	7.88	0.55	4.8	8.78	16.99	104.0	6.12	2.52	8.6	3.39
Daytona Beach	1250	1.09	7.2	6.65	0.88	7.2	8.23	0.52	4.8	9.29	8.18	59.9	7.32	2.3	8.6	3.72
Duke University	1023	-----	-----	-----	-----	-----	-----	-----	-----	-----	5.88	57.0	9.69	1.75	8.6	4.89
Fort Valley State	1110	0.73	7.2	9.92	0.59	7.2	12.28	0.34	4.8	14.21	7.18	69.8	9.72	2.33	8.6	3.67
Kentucky Adv Tech	1300	0.86	7.2	8.42	0.70	7.2	10.35	0.43	4.8	11.23	6.35	55.6	8.75	2.13	8.6	4.01
Kentucky Tech	1595	1.03	7.2	7.03	0.96	7.2	7.55	0.59	4.8	8.19	16.22	105.5	6.50	-----	-----	-----
Louisiana Tech	1695	1.38	7.2	5.25	1.18	7.2	6.14	0.63	4.8	7.67	18.58	96.9	5.22	2.53	8.6	3.38
University Cent. Fl.	1703	1.01	7.2	7.17	0.90	7.2	8.05	0.60	4.8	8.05	18.36	121.1	6.60	1.75	8.6	4.89
				Avg. 7.28			Avg. 8.64			Avg. 9.63			Avg. 7.22			Avg. 3.90

Phoenix: APS Solar & Electric 500, March 1993

Efficiency Event (const. speed)

School	Mass (kg)	Init. B. (V)	Fin. B. (V)	Drop (V)	Dist. (km)	Energy (kWh)	Effic. km/kWh
Cactus	1487	103.0	97.0	6.0	37	6.88	5.38
Camelback	1645	102.0	97.5	4.5	27	5.37	5.10
Carl Hayden #13	1459	104.5	102.0	2.5	23	3.54	6.37
Carl Hayden #69	1349	104.0	90.0	14.0	47	7.44	6.28
Chapparral	1481	104.0	90.4	13.6	69	8.93	7.75
Cortez	1415	105.0	90.0	15.0	76	10.64	7.11
E. Valley Institute	1481	103.5	88.0	15.5	68	10.45	6.47
Farmington	1471	102.5	87.5	15.0	52	10.75	4.79
Holbrook	1367	104.0	86.0	18.0	66	9.89	6.67
Marcos de Niza	1580	103.0	94.0	9.0	45	7.20	6.26
Marcos de Niza #93	1509	104.5	87.5	17.0	53	8.80	6.04
Marcos de Niza #94	1451	104.5	91.0	13.5	48	8.27	5.84
McClintock	1323	104.5	88.5	16.0	40	6.31	6.38
Mountain View	1345	103.5	99.5	4.0	21	3.12	6.71
Page	1462	103.5	89.5	14.0	82	13.00	6.32
Palo Verde	1517	108.5	91.0	17.5	56	11.39	4.95
Paradise Valley	1464	102.5	85.0	17.5	71	9.19	7.71
St. Johns	1334	105.5	91.0	14.5	76	9.85	7.68
Shadow Mountain	1307	103.0	95.0	8.0	48	8.21	5.88
South Mountain	1426	103.5	92.5	11.0	47	7.76	6.02
Sunnyside	1533	104.5	92.5	12.0	64	10.21	6.31
Window Rock	1409	104.0	98.5	5.5	21	3.56	5.88
							Avg. 6.27

Dearborn: 1993 HEV Challenge, June 1993

		(const. speed) DPG ZEV			(min. speed) MIS ZEV			(mixture) Total ZEV		
School	Mass (kg)	Energy (kWh)	Dist. (km)	Effic. km/kWh	Energy (kWh)	Dist. (km)	Effic. km/kWh	Energy (kWh)	Dist. (km)	Effic. km/kWh
Ground-Up Class										
Cal Poly Pomona	1248	5.06	33.2	6.55	----	----	----	5.06	33.2	6.55
Cal Poly SLO	1223	3.70	25.1	6.79	----	----	----	3.70	25.1	6.79
Cornell University	1155	3.98	33.2	8.33	5.93	48.3	8.15	9.91	81.5	8.22
Lawrence Tech.	1650	3.98	16.7	4.21	3.82	25.8	6.74	7.80	42.5	5.45
Michigan State	1478	4.60	33.2	7.21	2.82	25.8	9.13	7.42	58.9	7.94
UC Davis	1062	3.81	37.7	9.89	6.41	48.3	7.54	10.22	86.0	8.41
UC Santa Barbara	1401	4.97	33.2	6.67	3.20	16.1	5.03	8.17	49.3	6.03
Univ. of Tenessee	1233	3.95	33.2	8.40	----	----	----	3.95	33.2	8.40
Conversion Class										
Col. School of Mines	1614	3.42	33.2	9.70	6.13	48.3	7.88	9.55	81.5	8.53
Col. State University	1722	4.94	33.2	6.71	----	----	----	4.94	33.2	6.71
Concordia University	1671	4.35	33.2	7.62	----	----	----	4.35	33.2	7.62
Jordan College	1694	4.41	33.2	7.52	----	----	----	4.41	33.2	7.52
Seattle University	1715	3.99	33.2	8.31	2.92	19.3	6.62	6.91	52.5	7.60
Stanford University	1660	5.59	33.2	5.93	2.08	12.9	6.19	7.67	46.0	6.00
U.S. Naval Academy	1717	5.09	33.2	6.52	----	----	----	5.09	33.2	6.52
University of Alberta	1633	4.21	33.2	7.88	4.64	54.7	11.80	8.85	87.9	9.93
University of Illinois	1643	3.91	33.2	8.48	----	----	----	3.91	33.2	8.48
Wayne State	1848	5.40	33.2	6.14	1.59	9.7	6.08	6.99	42.8	6.13
Weber State	1725	4.29	33.2	7.73	3.25	29.0	8.92	7.54	62.1	8.24
				Avg. 7.40			Avg. 7.64			Avg. 7.43

Table A7. Representative 1993 American Tour de Sol and Atlanta vehicle data comparison

Vehicle Name	Mass (kg)	Vehicle Model	Energy (kWh)	Dist. (km)	Leg Time (hrs)	Avg. Speed (km/h)	Energy Economy (km/kWh)	Battery Type	Motor Manuf.	Motor Type	Controller
Univ. of Central Florida *	1,710	Mercury Lynx	18.36	120.4	2	60.2	6.56	Pb-Acid	G.E.	DC Series	G.E.
Kineticar	1,530	VW Rabbit	17.77	139.2	1.87	55.2**	7.83	Pb-Acid	Adv. DC	DC Series	Curtis
Electric Jewel	1,272	Geo Metro	13.04	156.8	1.96	52.7**	12.1	Pb-Acid	Prestolite	DC Series	Curtis
T-Star	1,125	Geo Metro	10.27	227.2	1.89	54.6**	22.26	Zn-Br	Solectria	AC Induction	Solectria

* from the Atlanta competition
** Based on the American Tour de Sol day 4 for the leg length of 103 km

Demco, Inc. 38-293

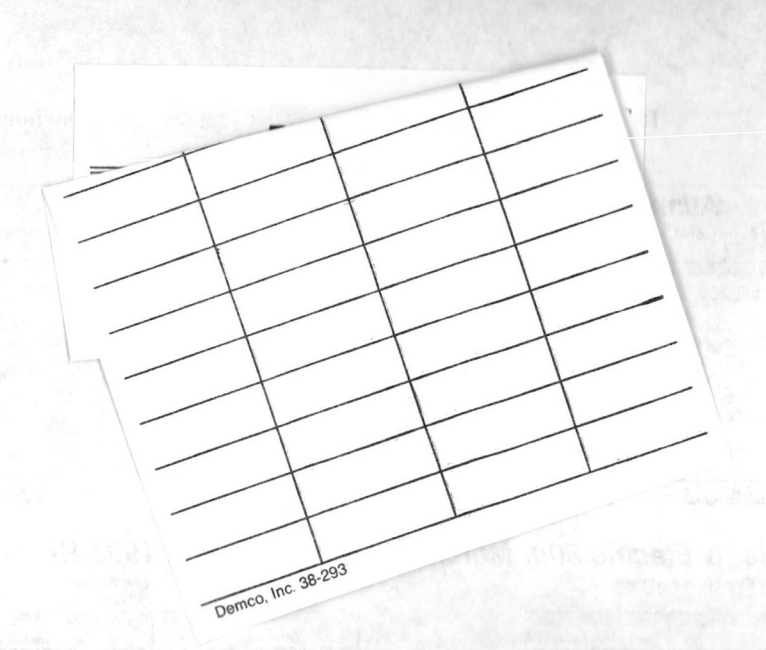